Evaluation of Cellular Processes by *In Vitro* Assays

Authored by

Taseen Gul

Ehtishamul Haq

&

Henah Mehraj Balkhi

Department of Biotechnology, Science Block, University of Kashmir, Hazratbal, Srinagar, India

Evaluation of Cellular Processes by *In Vitro* Assays

Authors: Taseen Gul, Mehraj Balkhi and Ehtishamul Haq

ISBN (Online): 978-1-68108-703-0

ISBN (Print): 978-1-68108-704-7

General:

1. Any dispute or claim arising out of or in connection with this License Agreement or the Work (including non-contractual disputes or claims) will be governed by and construed in accordance with the laws of the U.A.E. as applied in the Emirate of Dubai. Each party agrees that the courts of the Emirate of Dubai shall have exclusive jurisdiction to settle any dispute or claim arising out of or in connection with this License Agreement or the Work (including non-contractual disputes or claims).
2. Your rights under this License Agreement will automatically terminate without notice and without the need for a court order if at any point you breach any terms of this License Agreement. In no event will any delay or failure by Bentham Science Publishers in enforcing your compliance with this License Agreement constitute a waiver of any of its rights.
3. You acknowledge that you have read this License Agreement, and agree to be bound by its terms and conditions. To the extent that any other terms and conditions presented on any website of Bentham Science Publishers conflict with, or are inconsistent with, the terms and conditions set out in this License Agreement, you acknowledge that the terms and conditions set out in this License Agreement shall prevail.

Bentham Science Publishers Ltd.
Executive Suite Y - 2
PO Box 7917, Saif Zone
Sharjah, U.A.E.
Email: subscriptions@benthamscience.org

BENTHAM SCIENCE

CONTENTS

FOREWORD

This book provides hands on information about the various *in vitro* techniques and is preceded by the information about the principles and the basics of the technique. The authors have clearly and concisely discussed several concepts and methods which will be useful for the under-graduate and post-graduate students during their laboratory courses. The manuscript will equally contribute to the day-to-day research work of the scholars.

I strongly recommend publishing of this book and I wish best of luck to all the authors.

This book provides hands on information about the various *in vitro* techniques and is preceded by the information about the principle and the basics of the technique. The authors have clearly and concisely discussed several concepts and methods which will be useful to the under-graduate and post-graduate students during their laboratory courses. The manuscript will equally contribute to the day-to-day research work of the scholars.

I strongly recommend publishing of this book and I wish best of luck to all the authors.

Dr. Khalid Majid Fazili
Department of Biotechnology
University of Kashmir,
Hazratbal, Srinagar,
India

PREFACE

During the course of my doctoral degree, I was very much interested in the methods involved in the understanding effect of drugs in *in vitro* conditions. The students and scholars often find it confusing to select a particular assay on cell viability, cell proliferation and apoptosis. While teaching at the post-graduate level, I got to know about the problems our students face in understanding the basics of *in vitro* assays. Keeping in view these facts, I tried to address these concerns by providing a basic introduction about various assays in the form of this book. The book gives comparative analysis of different assays and discusses the advantages and disadvantages related to it. Each experimental protocol is preceded by the information about the principle and the basics of the technique.

I hope my little effort will benefit the students and scholars who are interested to know the basic fundamentals of cellular assays. To make this book better, the criticism and suggestion are most welcomed.

Taseen Gul
Department of Biotechnology,
University of Kashmir,
Hazratbal, Srinagar,
India

ABBREVIATIONS

ABBREVIATIONS

BAC Bacterial Artificial Chromosome

Bp Base pair

BrdU 5-bromo-2'-deoxyuridine

cDNA Complementary DNA

CFSE Carboxy fluorescein diacetate succinimidyl ester

ChIP Chromatin Immunoprecipitation

DAPI 4,6-Diamidino- 2-phenylindole

DIG Digoxigenin

DMEM Dulbecco's Minimal Essential Media

DMSO Dimethyl sulphoxide

DNTB 5-5'-Dithiobis [2-nitrobenzoic acid]

ELISA Enzyme Linked Immunosorbent Assay

EMSA Electrophoretic Mobility Shift Assay

FITC Fluorescein Isothiocyanate

FBS Fetal Bovine Serum

HRP Horseradish peroxidise

IC Inhibitory Concentration

IL Interleukin

JC-1 5,5',6,6'-tetrachloro- 1,1',3,3' tetraethylbenzimidazolcarbocyanine iodide

KDa Kilo Dalton

LDH Lactate Dehydrogenase

mRNA messenger RNA

miRNA micro RNA

MDA Malondialdehyde

NAD+ Nicotinamide adenine dinucleotide

NO_3^- Nitrate

NO_2^- Nitrite

NO Nitric oxide

PAGE Polyacrylamide Gel Electrophoresis

PARP Poly ADP-ribose polymerase

PI Propidium iodide

RB Retinoblastoma

ROS   Reactive oxygen species

RT-PCR   Real time polymerase chain reaction

T-DNA   Transfer DNA

TNF-α   Tumor necrosis factor alpha.

TNFR1   Tumor necrosis factor receptor 1

TUNEL   Terminal deoxynucleotidyl transferase dUTP nick end labelling

TBA   Thiobarbituric acid

uM   micromolar

WST   {2-(2-methoxy-4-nitrophenyl)-3-(4-nitrophenyl)-5-(2,4-disulfophenyl)-2H-tetrazolium}

XTT   2,3-bis-(2-methoxy-4-nitro-5-sulfophenyl)-2H-tetrazolium-5-carboxanilide.

CHAPTER 1

Basics of *In Vitro* Cell Culture

Abstract: In order to get the intricate biology of living organisms, the researchers started using the basic unit of life *i.e.* "Cell" for elucidating the intricate mechanisms. This led to the basis of cell culture studies, where cells grown in controlled artificial conditions simulate the conditions prevailing in natural ones, therefore, presumed to act as those in *in vivo* conditions. The introduction of cell culture techniques has helped a lot in understanding the physiological processes like cell signalling, neurobiology, cell proliferation, pathogenesis of diseases, apoptosis and even more. In the following chapter, we will discuss the basics of cell culture including the equipments, chemicals and different types of materials required for cell culture. The techniques used for maintenance, preservation and authentication of cell lines are also included.

Keywords: Antibiotics, Aseptic, Cell Culture, Cell Lines, Cryopreservation, Dimethyl-sulphoxide, Dulbecco's Minimal Essential Media, Fetal Bovine Serum, Fibroblast, Glutamine, Hood, Incubator, Media, Phase Contrast Microscope, Trypsin.

INTRODUCTION

Characteristic feature of all animals is that they are multi-cellular, consisting of many different cells, with specialized functions. Basic metabolic pathways in different cells are same with similar organelles however each cell also has a specific function and thus have unique expression of some of these components to perform specific function within the organism. Depending on specific roles, each cell type has a specific gene and protein expression with a characteristic shape, size, structure and function. They are said to have differentiated and are highly organised structures with specialised functions. The science and technological interventions have done marvels in understanding the cellular phenomenon and allowed us to grow cells even under artificial conditions. Isolated cells, tissues or organs can be grown in plastic dishes when they are kept at defined temperatures using an incubator and supplemented with a medium containing cell nutrients and growth factors.

First detailed study on *in vitro* culture of cells was made by Jolly in 1903. He studied cell survival and cell division in leukocytes of a salamander. Earlier during such studies, experiments were conducted on chunks of tissues and as such the name 'tissue culture' was used. However now, the term tissue culture is used for *in vitro* culture of cells only when cell culture is kept for more than 24 hours. In the evolution of animal cell culture studies, the major breakthrough was done by Ross Harrison, whose fundamental work in animal cell culture is considered as a corner stone of animal cell culture in science [1]. In 1907 he experimented on clotted lymph fluid using explants of frog embryo and observed cell proliferation in a depression slide. This technique continued to be used since then. It involves suspension of a drop over a depression in a microscopic slide sealed by a cover slip. This technique was further developed by Burrows with the use of plasma clots. Although they could evolve the technique significantly but the major hassle in the cell culture was the maintenance of the cultures without contamination from bacterial or other microbial cells. Since bacterial cells grew at a very fast rate in comparison to growth rates of animal cells, even a low-level contamination lead to unacceptable rate of contamination. To overcome this problem a surgeon Alexis Carrel, introduced strict aseptic techniques for cell culture *in vitro*. In order to sustain aseptic conditions for cell culture he introduced the 'Carrel flask' for aseptic subculture of cells and thus became the primogenitor of modern tissue culture. However, it was a lengthy procedure and difficult to repeat creating hurdles to adopt cell culture as a routine laboratory technique. In 1912, the first cell culture for a long period of 34 years was successfully carried out by Carrel, who started to culture chick embryo heart cells. This led to the imprecise belief that cells could be cultured for an indefinite period if given the appropriate conditions. Later it was observed that cell growth was maintained by the use of embryo extracts, but new cells were being continuously added to the culture during medium replenishment. In 1961, Hayflick and Moorhead established the finite capacity for cell growth. This was followed by a significant advancement in culturing technique by introduction of Trypsinization, by Rous and Jones in 1916. Trypsinisation involved the treatment of cells by the proteolytic enzyme trypsin to free cells from tissue matrix. Tissue culture contains a mixture of cell types and to produce homogeneous cell strains from such tissue cultures trypsinization was a breakthrough marking the start of animal cell culture techniques. Table **1.1** represents the major events that took place in the history of advancements in the field of culture techniques [2].

Table 1.1. Major Breakthrough in the History of Cell Culture.

1885	Cultivation of embryonic chick cells in a saline culture.	**Roux**
1897	Maintenance of blood and connective tissue culture in serum and plasma.	**Loeb**
1903	Study of cell division of salamander leucocytes *in vitro*.	**Jolly**
1907	Study and cultivation of frog nerve cells in a lymph clot *in vitro*.	**Harrison**

(Table 1.1) contd.....

1910	Study of mitosis and cultivation of chicken embryo cell in plasma clots.	**Burrows**
1911	Preparation of first liquid media.	**Lewis and Lewis**
1913	Introduction of aseptic techniques for cell culture.	**Carrel**
1916	Introduction of trypsin for the subculture of adherent cells.	**Rous and Jones**
1923	Development of 'Carrel' or T-flask as the first cell culture vessel.	**Carrel and Baker**
1927	Production of the first viral vaccine.	**Carrel and Rivera**
1933	Development of the roller tube technique.	**Gey**
1948	Isolation of mouse L fibroblasts. Development of chemically defined medium, CMRL 1066.	**Earle Fischer**
1949	Cultivation of polio virus on human embryonic cells in culture.	**Enders**
1952	Establishment of a continuous cell line from a human cervical carcinoma known as HeLa (Henrietta Lacks) cells. Developed plaque assay for animal viruses.	**Gey Dulbecco**
	Discovery of contact inhibition.	**Abercrombie**
1955	Studied the nutrient requirements of selected cells in culture and established the first widely used chemically defined medium.	**Eagle**
1961	Isolation of human fibroblasts (WI-38).	**Hay flick and Moorhead**
1964	Introduction of the HAT medium for cell selection.	**Littlefield**
1965	Introduction of the first serum-free medium.	**Ham**
1965	Fusion of human and mouse cells by the use of a virus.	**Harris and Watkins**
1975	Establishment of hybridoma for secretion of monoclonal antibodies.	**Kohler and Milstein**
1978	Development of serum-free media from cocktails of hormones and growth factors.	**Sato**
1982	First recombinant protein human insulin licensed as a therapeutic agent.	
1985	Human growth hormone produced from recombinant bacteria accepted for therapeutic use.	
1986	Lymphoblastoid γIFN licensed.	
1987	Tissue-type plasminogen activator (tPA) from recombinant animal cells commercialized.	
1989	Recombinant erythropoietin in trial.	
1990	Recombinant products in clinical trial (HBsAG, factor VIII, HIVgp120, CD4, GM-CSF, EGF, mAbs, IL-2).	

1.1. Tissue Culture and Its Types

In vitro culture of cells, tissues and organs is commonly referred as tissue culture. Tissue culture is cultivation of animal as well as plant cells *in vitro*. Tissue culture is classified into three major groups; organ culture, explant culture, and cell culture [3].

Organ Culture: The three-dimensional culture of tissues so that some or all of the histological features of the tissues are retained. The organ culture is carried out in such a way that cells under cultivation differentiate in a proper architecture. Organs cultured are able to accurately function in various states and conditions as actual *in vitro* organ itself. Organ culture could be carried out through different methods as:

Plasma Clot Method: A simple method for organ culture is to begin with embryonic cells rather than cells of adult origin. One of the methods employed for such a culture is plasma clot. In this method watch glass is placed over a pad of moist cotton wool in a petridish and drops of plasma are mixed with embryo extract in a watch glass in the ratio of 3:1. Dissected small piece of tissue and a raft of lens paper or rayon net are placed on top of the plasma clots in watch glass. The transfer of tissue occurs easily by raft. After removing the excessive fluid, the net with the tissue is placed again on the fresh pool of medium.

Agar Gel Method: Agar is a jelly like substance derived from polysaccharide agarose present in cell walls of some algae's. It is released on boiling and solidifies on cooling. Agar is used for growth and differentiation of organs from embryonic cells as they grow well on agar. Enriched agar solidified with defined media with or without serum is also used for organ culture. These media consist of 1% agar in Basal Salt Solution (7 parts), chick embryo extracts (3 parts) and horse serum (3 parts). The medium with agar provides the mechanical as well as nutritional support for organ culture. An adult organ culture requires more oxygen and thus is more difficult to cultivate. A special equipment known as Towell's II culture chamber has been used to study variety of adult organs using special media. This particular chamber permits the use of 95% Oxygen.

Raft Method: This method involves the use of raft of lens paper or rayon acetate to place the explants, which float on serum in a watch glass. The rafts are treated with silicone which enhances floatability of lens paper. The combination of clot and raft techniques involves the explants to be first placed on a suitable raft and then on a plasma clot. This modification makes media changes easy, and prevents the sinking of explants into liquefied plasma

Grid Method: This method was designed by Trowell in 1954. It involves the use of a wire mesh or perforated stainless steel sheet of 25 mm x 25 mm and their edges are bent to form 4 legs of about 4 mm height. The tissues are either placed directly on the grid like skeletal tissue whereas the softer tissues like glands or skin are first placed on rafts and then on grids. The grids are then placed in a chamber filled with fluid medium and supplied with a mixture of Oxygen and Carbon dioxide to meet the Oxygen requirements of adult mammalian organs. Several modifications are done in this method to study the growth and differentiation of adult and embryonic tissues.

Explant or Organotypic Culture: In Explant culture, an appropriate substrate is used to attach small pieces of the tissue and is cultured in serum rich medium. Earlier, explants were maintained on a depression in a thick glass slide known as Maximov chambers. Nowadays, much more convenient method is used which employs use of regular culture dishes. The technique requires high expertise as explant cultures if handled properly can be maintained for long time, and cells within the explant continue to differentiate properly. The tissue's architecture is preserved within the explants and is the principal advantage of this method.

Dissociated Cell Culture: Cultures derived from dissociated cells taken from an original tissue is known as dissociated cell culture or simply cell culture. Cells are first disaggregated either mechanically or enzymatically and cells obtained are cultured as a monolayer on a solid substrate, or as a suspension in the culture medium. They can be propagated and hence expanded and divided to give rise to replicate cultures. One of the advantages of cell culture is that it makes each living cell accessible. Dissociated cell cultures are best to study morphological and physiological techniques, which can be applied on a cell by cell basis. One of the limitations is to maintain conditions that permit good cell growth and maturation. Cells in dissociated cultures often lose histotypic architecture as well as some of the biochemical properties associated with it.

1.2. Cell Line

Subculture of the primary culture leads to the generation of cells known as a sub clones or cell lines. These sub clones or cell lines have a limited life span and on passaging cells with the highest growth capacity predominate, resulting in a degree of genotypic and phenotypic uniformity of the predominating cell population. When a subpopulation of a cell line is passaged further and subjected to assortment, the cell line selected through a positive selection is known as a cell strain. Each cell strain has specific peculiarities; morphological as well as genetic different from parental cells as well as other sub population of cells.

Normal cells can be sub cultured only a limited number of times as they have a limited capacity to divide, these cells are known as finite cell lines. Finite cell lines undergo a genetically determined process known as senescence. However, some cell lines can divide infinite number of times without undergoing the process of senescence. Such cells are immortalized through a process known as transformation. The process of transformation can occur spontaneously or can be induced chemically or virally. If a finite cell line undergoes transformation and acquires the ability to divide indefinitely, then it is termed as a continuous cell line.

1.3. Types of Culture

Primary Cell Culture: Involves direct culturing of cells taken from the tissues of an organism and maintaining it in growth medium under aseptic conditions. Primary cells have a finite life span, undergo the process of senescence and consist of a heterogeneous mixture of cells.

Secondary Cell Culture: It is prepared from the primary culture. It refers to the transfer of cells from one culture vessel to the other. Sub culturing of primary cells leads to the generation of cell lines.

Adherent Culture: Cells are of two types either anchorage dependent or anchorage independent. The cells which are anchorage-dependent are cultured on a suitable substrate that allows cell adhesion and spreading. Except haematopoietic cells and few others, the majority of the cells derived from vertebrates are anchorage-dependent.

Suspension Culture: Cell lines which are adapted for anchorage independent culture form suspension culture. These cells are cultured in dishes or flasks which are not treated for adherence. To provide adequate gas exchange medium is agitated. This agitation is usually achieved with a magnetic stirrer or rotating spinner flasks.

Cells which are differentiated have limited ability to proliferate as such they do not contribute to the formation of a primary culture. Differentiated cells require special conditions for attachment and preservation of their differentiated status. In a tissue, a part of cells are committed to proliferate and represent the largest number of proliferating cells, such as the fibroblasts of the dermis. Tissues also contain a small population of regenerative cells which when under suitable conditions may also undergo culture, which may be propagated either as stem cells or lead to differentiation. This implies that if we want to retain the proliferative capacity of cells, we must choose the correct population of cells and suitable conditions for expansion of the cell population. Such conditions were mimicked in culture of fibroblastic cells by the inclusion of factors in serum that help to maintain the proliferative precursor phenotype of cells. Beside different sets of conditions required, may be used, one for proliferation and one for differentiation as per the requirements. Not only selective media but the correct matrix interaction, cell-cell interaction and, cellular polarity is required to be established. For carrying out extensive tissue repair or replacement the graft needs to be fully or partially differentiated.

Currently, the cell culture techniques have become cornerstone tools for cell and molecular biology and are widely used as;

1. Model systems for understanding cell biology-interactions, pathways, aging *etc*.
2. Toxicity testing of new drugs
3. Cancer research by studying the effect of chemicals, radiation and virus on cell lines.
4. Production of vaccines from virus and to explore their lifecycles and pathogenesis.
5. Gene therapy.
6. Production of monoclonal antibodies, insulin and hormones *etc*.

The culturing of cells is a sophisticated approach and requires a number of pre-requisites. It requires a separate place to establish culture lab which would have minimal interference of the people. The key requirements include the specialised equipments, cell culture grade chemicals, high quality plastic ware *etc*. All of them are discussed in the following sections [4].

1.4. Equipments for Cell Culture Lab

The essential requirements for the development of cell culture facility are:

Hood: The sterile environment is one of the essential features for the culturing of cells. Laminar hoods provide aseptic environment for the manipulation of cells, cell culture vessels and solutions *etc*. The access of microorganisms is restricted by the forceful air flow of the laminar hoods. Hoods can be either horizontal or vertical depending upon the direction of the air flow. They are provided with UV light fixtures so as to sterilise the work area and illuminated few minutes prior to work. There should be lowest possible items inside the hood so as to avoid chances of contamination (Fig. **1.1**). All the bottle tops and necks should be wiped with 70% ethanol and the spills in the hood should be cleaned up immediately [5].

Incubator: Maintaining temperature is one of the essential pre-requisite for the growth of cells and it is provided by the Incubators. It is a simple thermostat CO_2 incubator with removable shelves and a water pan at the bottom for maintaining humidity. Usually a temperature of 25-30°C for insect cell line and 37°C for mammalian cell line is preferred. The incubators should be properly maintained and fumigated periodically with bleaching agents.

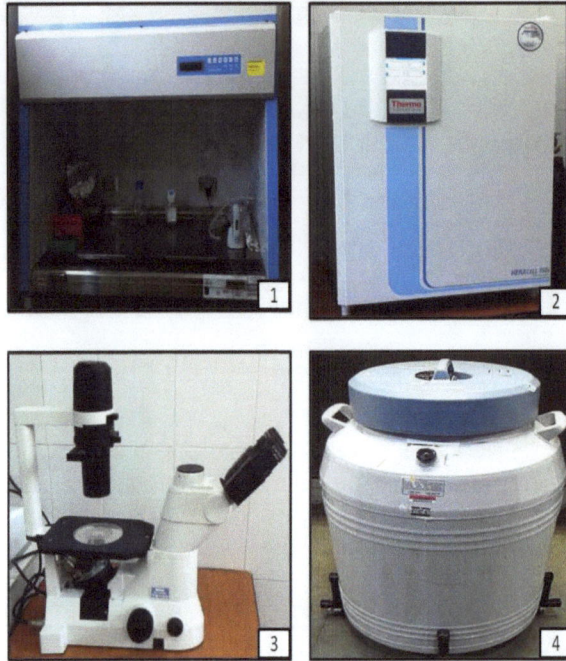

Fig. (1.1). Different equipments used in the cell culture lab.
(1) Hood, (2) Incubator, (3) Phase contrast microscope, (4) Storage tanks.

Microscope: For the monitoring of growth, effect and contamination of cell lines, microscopes are necessary. Usually, inverted microscope with 10X-100X magnification is used.

Water Bath: The water baths are used to warm the cell culture media, trypsin and thawing freeze downs prior to use. They are usually maintained at a temperature of 37°C. They are the principal source of contaminations and should be looked after carefully. The autoclaved water should be used and frequently changed.

Refrigerators, Centrifuges, Autoclaves and Storage Tanks: The media, enzymes, antibiotics, serum, buffers *etc*. should be kept in refrigerators at 4°C. Separate bench top centrifuges should be used for cell culture. The sterilization of water, buffers and some of cell culture plastic ware should be carried out in autoclaves. However, 0.2 micron filters should be used for the filtration of heat labile chemicals. The liquid nitrogen storage tanks or -80°C freezers are used for the cryopreservation of cell lines.

Cell Culture Plastic Ware: The cell culture plastic ware should be of high quality grade. The flasks, dishes and assay plates, pipettes *etc.* are of different sizes and capacities and are used depending upon the requirement as shown in Fig. (**1.2**). The pipette aid used for introducing media in and out of the culture vessels should be used carefully and kept inside the hood so as to avoid cross contamination from other cell lines. Individually wrapped glass pipettes are best for cell culture [6]. The plastic ware should be immediately disposed off after use.

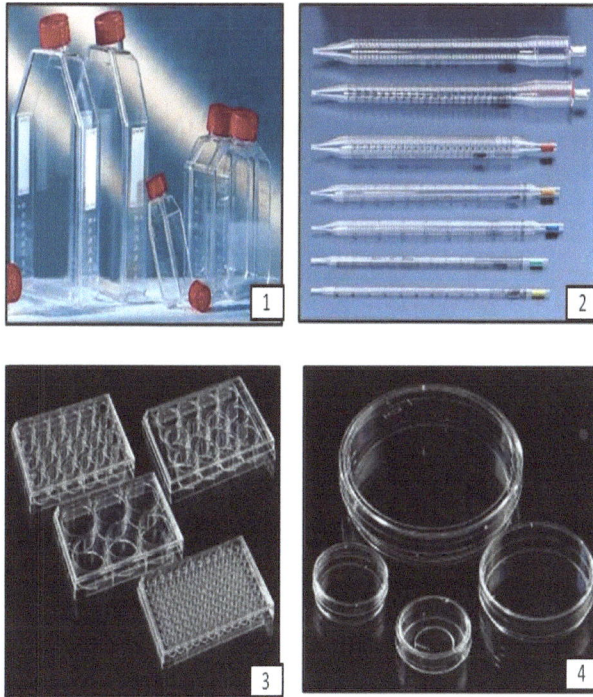

Fig. (1.2). Different types of plastic ware used in a cell culture lab.
(1) Flasks of different areas ($25cm^2$, $50cm^2$, $75cm^2$, $150cm^2$), (2) Pipettes of different volumes (2ml, 5ml, 10ml, 25ml, 50ml), (3) Culture plates (6well, 12 well, 24 well, 96 well), (4) Culture dishes (3cm, 6cm, 10cm *etc.*).

1.5. Materials for Cell Culture

Cell culture is a highly sophisticated technique and a lot of materials are needed for standardisation and finally establishing a cell culture lab. Following are the diverse materials required to grow cells under artificial conditions.

Media: The culture media is a complex mixture of carbon and energy source, serum, vitamins, minerals, growth factors, trace elements and buffers in water. The growth medium contains components that incorporate into dividing cells and maintain the proper chemical environment of the cell. In 1955, Harry Eagle first described the media composition. The media requirements are specific for the cell line used and are available in both powdered and liquid form. Carbohydrates are the major sources of energy and among them glucose is the most frequently used sugar (0.8-5g/litre). The other sugars such as sucrose, mannose, galactose, fructose *etc.* may also be included. The animal cells require essential amino acids like arginine, cysteine, isoleucine, histidine, lysine, phenylalanine, threonine, tryptophan and valine *etc*. Glutamine is also one of the essential requirements for the growth of cells. It acts both as energy source and in the synthesis of nucleic acids. Some of the amino acids are added to compensate specially for a particular cell type. The vitamins which are most commonly added to the culture growth media include biotin, choline, nicotinic acid, folic acid, pantothenic acid, riboflavin, thiamine, para-aminobenzoic acid, thiamine pyridoxal and inositol. Hormones and growth factors also play an important role in the survival and proliferation of cells. Insulin and hydrocortisone are the most widely used whereas interleukins, epidermal growth factor and non-epidermal growth factor are the less frequently used. The osmolality of the solution is usually maintained by amino acids, glucose and ions. The media also contains phenol red as pH indicator. The commonly used basal media formulations are Dulbecco's Modified Eagles Media (DMEM), Ham's F12, RPMI 1640 *etc*.

Serum is an essential part of complete growth media and usually 5-20% of serum is added in the growth media. Serum is composed of a cocktail of growth factors and thus serves as a universal growth supplement. Though the exact composition of serum is not known but it has several functions. It stimulates cell growth due to the presence of growth factors, enhances cell attachment and provides transport proteins (hormones, minerals and lipids). Usually, Fetal Bovine Serum, Fetal Calf Serum and Horse Serum are used. However, careful selection and validation of serum sources should be adopted.

Trypsin: It is an enzyme used to segregate the cells from the culture vessels. It's used for adherent cultures only. Trypsinisation should be carried out carefully as excessive trypsinisation can lead to stress or even death of cells.

Antibiotics: In order to avoid bacterial or fungal contamination in the cell culture, antibiotics are added. Usually a combination of penicillin (100 IU/ml) and streptomycin (50ug/ml) is used for bacterial contamination and Gentamycin (50ug/ml) to overcome yeast contamination. The other anti-fungal agents include Amphotericin B (2.5ug/ml) and Nystatin (25ug/ml). The use of antibiotics should be avoided so as to reduce the microbial resistance issues.

Dimethyl Sulphoxide: It's a cryoprotectant chemical and is used for the storage of cell lines.

Phosphate Buffer Saline: PBS is the commonly used buffer for the cell culture. The buffer should be made regularly and autoclaved prior to use. It's used to wash the cells before trypsinisation.

1.6. Cryopreservation of Cell Lines

Cryopreservation has been derived from a Greek word "kruos" which translates to "frost" and Latin "praeservare" which means "to keep". Preservation of cells at a low temperature, below -100°C to maintain structurally intact living cells and tissues is known as cryopreservation. By cryopreservation a renewable source of cell line can be obtained so that experiments can be done over a period of time without incurring high costs and trouble of continuous culture [7].

Cryopreservation is a very elaborate process which involves following main steps:

Gentle Culture of Cells: Cryopreservation involves long term storage of cells under stringent conditions. Cells which are in an actively growing phase are used, to ensure maximum health and a good revival. Prior to preservation, cells are maintained in an enriched media and a day before preservation culture media is changed. Before processing the cells for cryopreservation morphological analysis of culture is carried out to examine the cellular peculiarities and general appearance of the culture. Further cell culture is checked for signs of microbial contaminations. Often such cultures which are to be cryopreserved are maintained in an antibiotic-free media so that any cryptic contaminants are detected.

Cell Harvesting: Harvesting is the most critical and determining step in the process of cryopreservation. In order to assure maximum viability, cells are harvested very gently to minimize the damage incurred during harvesting, freezing and thawing processes. At the time of harvesting, cells should neither be less confluent nor over confluent. Cells are harvested at near confluence density (1.5×10^7 cells/T-75ml flask). Ideally cells are cryopreserved at a cell density of 2×10^6 cells/vial. While harvesting, maintenance of sterility is of outmost importance so sterile pipettes are used throughout the process. Cell monolayer is thoroughly washed with calcium and magnesium free phosphate buffered saline to remove all traces of fetal bovine serum. After wash cell monolayer is treated with pre-warmed trypsin solution which leads to disintegration of cell monolayer. In order to avoid cell clumping a proper trypsin treatment is compulsory for rounding off the cells. When cells loosen up due to the trypsin treatment the progress of the enzyme action is observed under an inverted phase contrast microscope to avoid over trypsinization of cells. After trypsinization rounded up cells are detached by gently tapping the flask followed by neutralization of trypsin with addition of growth medium to the cell suspension. Cell suspension is vigorously pipette and placed on ice. Before processing the cells further, cell counting of the sample is carried out by Trypan blue method and number of cells/mL and the total cell number is calculated. Cells are pelleted by centrifugation at 100xg for 5 minutes to obtain a cell pellet. Centrifuged cells are resuspended in a special cryo-medium according to the determined cell density of the sample.

Freezing: Cooling below a certain temperature is typically accompanied by freezing of water and consequent concentration of the solute component. This leads to freezing injury in cells either by the direct mechanical action or from secondary effects due to changes in the liquid phase. To prevent freezing damage to cells a cryoprotectant media is used for suspending the cells for cryopreservation. Cryoprotectant permeabilizes plasma membrane so that water flows out of the cell increasing the total concentration of all solutes. The cells are dehydrated and ice crystals form in the surrounding medium and not inside the cell thus preventing the cellular damage.

Many chemical compounds have been used as cryoprotectant such as glycerol, dimethyl sulfoxide and ethanediol. A cryoprotectant should be able to penetrate into the cells without producing any toxic effects to cells as it is imperative not only to ensure cell survival but also to avoid damage to the cell structure. Among vast variety of cryoprotectants, dimethylsulfoxide (DMSO) and glycerol are the most widely used. DMSO is a very powerful solvent but potentially hazardous as it can penetrate intact skin. It is highly recommended to avoid contact with DMSO and to properly dispose of any wastes containing DMSO properly. DMSO is used in the cryopreservant media in a concentration range of 5 to 15% (v/v) depending upon the type of cell. Glycerol is generally used at 5 - 20% (v/v). Glycerol is less toxic to cells but often leads to some osmotic problems, especially after thawing. For highly sensitive cells standard cryoprotectant mixtures are replaced with serum rich media (95% serum + 5% DMSO) which helps sensitive cells survive freezing. Glycerol should be added at room temperature and removed slowly by dilution. To reduce the DMSO toxicity, lower concentrations should be used and added to the cell suspension at 4°C and removed immediately upon thawing.

After suspension in proper cryopreservant media cells are kept in freezing vials which are of high grade quality so that they can tolerate low temperatures. These vials are labelled appropriately with the essential information such as cell line, date and a unique serial against which appropriate information about the cells is maintained. Elaborate details such as culture's storage conditions, culture identity, cell passage number, doubling time, cell density, date frozen, freezing medium and method used are noted against the unique serial number. Supplementary information, such as its origin, history, growth parameters, special characteristics and applications are also included.

Initially vials are placed in the rate controlled freezers for 24 hours in a Styrofoam container. Rate of cooling should be carefully maintained depending upon cell type and size. Most efficient rate of cooling is -1°C to -3°C per minute. Cooling rate should be such that enough time is available to cells to dehydrate but fast enough to prevent excessive dehydration damage. Currently rate of cooling is controlled by using mechanical freezing units, electronic programmable freezing units or cooling racks. For slow freezing alcohol-filled containers are used to slowly freeze vials placed in the system. Alcohol acts as a bath to achieve uniform heat transfer. Vials are placed in it overnight, and finally transferred to their final storage locations.

For long term vials are stored at a temperature below -130°C. Liquid nitrogen cooled freezers are preferably used in culture laboratories. Liquid nitrogen freezers maintain temperature between -140°C and -180°C and allow storage either above the liquid in the vapour phase, or at a temperature below -196°C submerged in the liquid. To reduce the possibility of the vials filling with liquid nitrogen during extended storage vapour phase storage is mostly used as it greatly reduces the possibility of leaky vials or explosion of ampules. Vials filled with liquid nitrogen may explode violently upon removal from the freezer, thus safety equipment should be used while removing vials and ampules from liquid or vapour phase nitrogen freezers. A full-face shield, heavy gloves and lab coat should be used while handling the vials or ampules for protection against explosions.

Cell Thawing and Recovery: Cell thawing is another most critical and determining step in the process of cryopreservation. Vials are removed from the storage using appropriate safety equipment. Before thawing the cells a quick check is carried out to ensure correct culture and special requirements of the cells if any. Vials are thawed quickly and gently in warm water. In order to prevent the formation of damaging ice crystals within cells during rehydration and ensure maximum viability, cells are quickly thawed within 60 to 90 seconds at 37°C. Once thawed the cryoprotectant media is quickly removed from the cells as prolonged exposure to cryoprotective agent's damage cells. Cells are revived in a fresh rich culture media based on characteristics of the cells. Most cells recover normally within 6 to 8 hours of thawing [8].

1.7. Cell Line Authentication

A large number of different cell lines have been derived from humans and other organisms. The researchers should be careful while selecting an appropriate cell line for their research purposes [9]. The cell lines procured from international

(ATCC) or national repositories (NCCS) should be checked carefully so as to avoid the false positive results. The cell lines should be monitored for microbial contamination. The contaminated cell lines should not be propagated. The cells should also be checked for cross contamination by other cell lines. The rough method for authentication of cell lines is by their morphology as shown in Fig. (**1.3**). However, to validate a particular cell type, following testing procedures should be done.

Isoenzyme Profiling: It is one of the most important biochemical characteristics for the species variation. By knowing the mobilities of different isoenzymes from different cell line homogenates; one can determine the species origin of cell line. Special kits are available for the identification of different enzymes and they provide the detection of different enzyme reactions. More the number of enzymes studied, more likely is the cell line identified. Thus, isoenzymes profiling enables quick speciation of cell culture and is of great help in routine testing. However it's hard to achieve the unique identification of a cell line.

Cytogenetic Analysis: This method is used to determine the karyotype of species or cell line. It also detects changes in cell culture and presence of cross contamination between cell lines. Though there are many different methods for staining chromosomes like Giemsa, G11 and quinacrine but G-banding is the most widely used. In this method, treatment of chromosomes with trypsin and giemsa show banding pattern that is unique for chromosomes. In G-11 banding, giemsa stain is used at a different pH *i.e.* at pH 11 instead of pH 6.8 which elicits the differential staining of chromosomes. However, flourescent bands are produced due to the use of quinacrine. Q-banding is most commonly used for the analysis of Y chromosome which is most distinctive using this method. It's used for the confirmation of cell line but the method is complicated and requires high expertise.

Southern Blotting: The genomic DNA is isolated from the cell lines and then digested by specific restriction endonucleases, followed by the separation of fragments by agarose gel electrophoresis. The digested DNA is then transferred to the nylon membrane and then hybridised by specific labelled DNA probe. The signals are then compared to standard cell lines and thus cell line characterised.

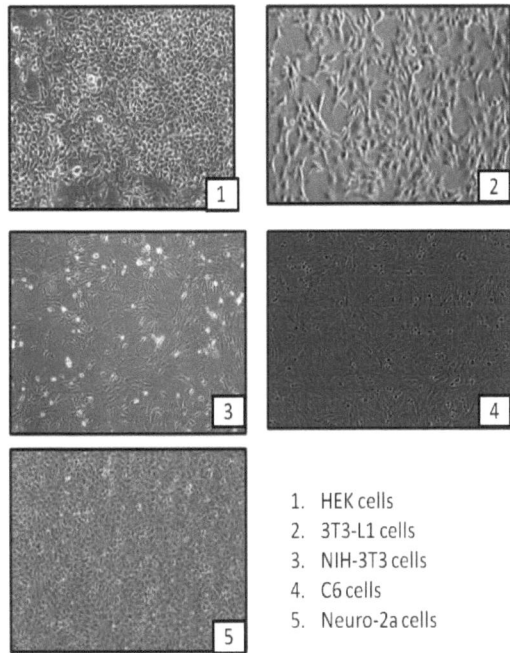

Fig. (1.3). Morphological characteristics of some commonly used cell lines.

1. HEK cells
2. 3T3-L1 cells
3. NIH-3T3 cells
4. C6 cells
5. Neuro-2a cells

Multiplex PCR: This method is an international reference standard for the cell line authentication. In this method, the STR loci are amplified by specific set of primers and the PCR products are analysed with size standards. The DNA samples from all species could be analysed by this technique but the validated primers should be available.

1.8. Contamination in Cell Lines

The transition for the application of the techniques of cell culture on a laboratory scale to large bio-processing units came with the capability of virus propagation in cell culture. The first breakthrough application was the production of polio vaccine in 1950's which was the first major commercial products of cultured animal cells. Since then a range of other beneficial products were synthesized from animal cells. Consequently, the study of the optimization of culture conditions to maintain consistently high productivity from such animal cells *in vitro* has become of utmost importance. Animal cell culture has become a promising alternative for animal experiments in the process of drug discovery and development. However, one of the major challenges for establishing cell cultures is to avoid the contamination in cell lines. The contamination if present can have disastrous effects not only on culturing of cells but also on the septic conditions of the entire cell culture lab. The contamination of cell cultures has very serious consequences in the present scenario [10]. Initially, the sources of contamination were the embryo extracts and animal blood serum which were added as growth supplements to the media for sustenance of cell culture. But these extracts had undefined composition and sometimes lead to contamination in cell lines. Then they were replaced by the nutrient formulations which had advantages of consistency between batches, ease of sterilization, and reduced chance of contamination.

Cell culture contaminants can be divided into two main categories,

Chemical Contaminants: these are the impurities which have their source in media, sera, and water, endotoxins, plasticizers, and detergents.

Biological Contaminants: these include contaminants such as bacteria, molds, yeasts, viruses, mycoplasma, as well as cross contamination by other cell lines.

Bacteria: Bacteria are ubiquitous and pervasive group of unicellular microorganisms. They are a few micrometers in diameter and have different morphologies, shapes and sizes. Because of their ubiquity, small size, and fast growth rates, bacteria are the most commonly confronted contaminants in cell culture. Bacterial growth can be visualized by simply inspecting the cultures; infected cultures usually appear turbid and there is a sudden drop in the pH of the culture medium. Bacteria appear as tiny moving granules when seen under a microscope whereas observation under a high power microscope can resolve the shapes of individual bacteria.

Yeasts: These are microscopic eukaryotic, unicellular microorganisms ranging in size from a few micrometers to 40 micrometers. The cultures contaminated with yeast appear turbid if the contamination is at an advanced stage. There is slight change in the pH of the cell cultures contaminated by yeasts until the contamination becomes heavy, at which stage the pH usually increases. Under microscope, yeast appears as individual ovoid or spherical particles that may bud off smaller particles.

Molds: are filamentous eukaryotic microorganisms. Their multicellular filaments contain genetically identical nuclei, and are referred to as a mycelium. The pH of cell cultures contaminated with molds remains stable in the initial stages of contamination, then rapidly increases as the culture become more heavily infected. Under microscopy, the mycelia usually appear as thin, whip-like filaments, and sometimes as denser clumps of spores. Spores of many mold species can survive extremely harsh environments in their dormant stage, only to become activated when they encounter favourable conditions.

Viruses: These microscopic infectious agents are extremely small in size that takes over the host cells machinery to reproduce. Most viruses have very stringent requirements and usually do not adversely affect the cell but they are very difficult to detect in culture. Virally infected cell cultures pose serious threat to the laboratory personnel. Viral infection of cell cultures is difficult to be detected with routine microscopy and can be detected only by electron microscopy, with a panel of antibodies or PCR with appropriate viral primers.

Mycoplasma: Mycoplasmas are the smallest self-replicating organisms, bacteria that lack a cell wall. There are often no visible signs of contamination in the cultures infected with mycoplasma. Mycoplasma are usually slow growing and persist in cell cultures without causing any cell death, but they can modify the behaviour and metabolism of the host cells in the culture. Mycoplasmas are very difficult to detect until they achieve extremely high densities. The only assured way of detecting mycoplasma contamination is by testing the cultures by using fluorescent staining like Hoechst 33258, ELISA or PCR,

Cross-Contamination: Contamination of cell cultures with other fast growing cell lines such as HeLa is a less frequent phenomenon but a problem with serious consequences. Cross contamination is not as common as microbial contaminations but has been clearly established. Cross contamination can be avoided or prevented by obtaining cell lines from only reputable cell banks and through a periodical checking of the characteristics of the cell lines in culture. Cross-contamination in cell cultures can be confirmed by the presence or absence of.DNA fingerprinting, karyotype analysis, and isotype analysis [11].

Methods to avoid contamination in cell culture:

It is impossible to eliminate contamination wholly; however it is possible to reduce its recurrence and seriousness so that the consequential damage is reduced. To minimise the chances of contamination, the following suggestions are recommended:

a. Do not use hood as a storage area. Keeping unnecessary bottles, cans, boxes *etc.* in the hood not only creates clutter but also disrupts the airflow patterns.
b. Mouth-pipetting is one the major contributing factors to contamination and should be completely avoided.
c. We should use clean lab coats specifically for cell culture purposes to avoid chances of shedding contaminations from the clothes. The hygiene of the individual involved in culturing is important factor to be considered.
d. Use sealed culture vessels to make it more difficult for micro-organisms to gain entry inside the vessels. The multi well plates can be sealed with labelling tape, smaller culture dishes can be placed in larger culture dishes or they can

be placed in sealable bags. The flasks with vented caps should be recommended for use. They have hydrophobic filter which allow gaseous exchange without allowing the passage of microorganisms.

e. Do not pour media directly from the large media bottles. We should use pipette aid for dispensing media, serum, *etc*. The disposable aspirators should be used for discarding the used media from the culture.

f. Work with a single cell line at a time and use of separate media bottles, pipettes and culture vessels *etc*. for each cell line to avoid the chances of cross-contamination.

g. The hood area, water baths, incubators *etc*. should be regularly monitored and treated with disinfectants as they can lead to mass contamination.

h. Wipe the working area, bottles, pipette aid *etc*. with 70% ethanol prior to use to kill the microorganisms.

i. Aliquot the 500ml, 250ml bottles of media, serum *etc*. into smaller volumes of 50ml, and 100 ml so as to losses arising due to contamination.

j. Periodically monitor the culture under microscope.

CONCLUDING REMARKS

The introduction of cell culture led to remarkable innovations like cell cloning, transgenics, tissue engineering and gene therapy. The cell culture studies involve the use of either primary cells or established cell lines. The primary cell culture involves the culture of tissues directly removed from living organisms. Primary cells closely resemble "real cells" but can be cultured only for few passages. However, an established cell line proliferates indefinitely through spontaneous or deliberate modifications. They have an ease of culture and are phenotypically more stable than primary cells. There are several established cell lines used extensively for cell culture assays. The *in vitro* culture techniques reduced the burden of using *in vivo* models for understanding biological mechanisms and to provide better comprehension of different cellular processes. The techniques used for large scale screening of drugs for different diseases, otherwise, would have been time consuming, highly expensive and burdensome.

CONSENT FOR PUBLICATION

Not applicable

CONFLICT OF INTEREST

None Declare

ACKNOWLEDGEMENTS

None Declare

REFERENCES

[1]   Harrison RG. The reaction of embryonic cells to solid structures. J Exp Zool 1914; 17: 521-44.
        [http://dx.doi.org/10.1002/jez.1400170403]

[2]   Ambrose CT. An amended history of tissue culture: Concerning Harrison, Burrows, Mall, and Carrel.
        J Med Biogr 2016; 1: 967772016685033.
        [PMID: 28092484]

[3]   Thomas J. Organ Culture. London: Elsevier: Academic Press 2012; pp. vii-viii.

[4]   Threfall G, Garland SG. In animal cell biotechnology. London: Academic Press 1985; pp. 123-40.

[5]   Block SS. Disinfection, sterilization and preservation. Philadelphia: Lea and Febinger Press 1983; pp.
        469-92.

[6]   Masters RW. Animal cell culture: a practical approach. Oxford University Press 2000; pp. 1-334.

[7]   Davis JM. Basic cell culture. Oxford University Press 2001; pp. 1-408.

[8]   Shannon JE, Macy ML. Tissue Culture: Methods and Applications. New York: Academic Press 1973;
        pp. 712-8.

[9]   Freshney RI. Animal cell culture: a practical approach. Washington, DC: RL Press 1986; pp. 242-71.

[10]   Fogh J. Contaminants in tissue culure. New York: Academic Press 1973; pp. 233-42.

[11]   MacLeod RA, Dirks WG, Matsuo Y, Kaufmann M, Milch H, Drexler HG. Widespread intraspecies
        cross-contamination of human tumor cell lines arising at source. Int J Cancer 1999; 83(4): 555-63.
        [http://dx.doi.org/10.1002/(SICI)1097-0215(19991112)83:4<555::AID-IJC19>3.0.CO;2-2]   [PMID:
        10508494]

CHAPTER 2

Methods of Transfection

Abstract: In order to evaluate the regulation, expression or activity of a gene, it is necessary to transfer the gene or its manipulated form into the *in vitro* systems. Since the mammalian cells do not uptake the foreign DNA efficiently, the availability of effective methods for introducing genes into the cells is essential. The transfection methodology has developed rapidly and diversely. Each year many new products and technologies are launched with improved efficiency and less cytotoxicity. In the following chapter, we will discuss the advantages and disadvantages of different methods of transfection.

Keywords: Cationic lipid, Diethylaminoethyl-dextran, Electroporation, Endocytosis, Genome, Gene transfer, Lipofectamine, Liposome, Micro-manipulator, Retrovirus, Transgene.

INTRODUCTION

Transfection is a method by which foreign DNA is introduced into the *in vitro* systems by chemical, physical or biological processes [1]. Transfection is of two types:

i. Stable transfection: the foreign DNA gets integrated into the genome.
ii. Transient transfection: the foreign DNA does not get integrated into the genome and is expressed for a limited period of time.

Table **2.1** represents compares the different aspects related to Stable and Transient Transfection.

Table 2.1. Comparison of Stable and Transient Transfection

S.No	Stable Transfection	Transient Transfection
1.	The transfected DNA gets integrated into the genome	Transfected DNA remains in the nucleus and does not get integrated into the genome
2.	Cells are harvested after 2-3 weeks as it first requires the selection of stably transfected colonies	Cells are harvested within 24-96 hours after transfection
3.	Selective screening is required for confirmation	Selection is not required

(Table 2.1) contd.....

S.No	Stable Transfection	Transient Transfection
4.	Only DNA vectors are used	Both DNA and RNA vectors can be used
5.	There is low level of protein expression due to low copy number of stably integrated genetic material	High level of protein expression due to high copy number of transfected genetic material
6.	Suitable for studies using vectors with inducible promoters	Not suitable for studies involving use of vectors with inducible promoters
7.	The transfected DNA stably passes from one generation to another, thus the genetic alteration is permanent	The genetic alteration is not permanent as the transfected DNA is not passed on to the progeny.

There are different methods of transfection but the choice of a method depends upon the following parameters *i.e.* for the optimal transfection method, following points should be taken into consideration.

- Type of delivered molecule.
- Quality and Quantity of DNA.
- Confluence of cells.
- Cell type.
- Cellular context.
- Transgene capacity.
- Safety issues.
- Length of expression.
- Expertise.
- Cost.
- Time.
- Desired efficiency.

Though the procedures for Transfection varies depending upon the above mentioned parameters but general procedure for Transfection is shown as a flowchart in Fig. (**2.1**).

The transfection techniques are commonly classified as [2]:

1. Biological methods which require use of genetically engineered viruses *e.g.* adenovirus, retrovirus *etc.*
2. Chemical methods which involve the use of carrier molecules for transfection *e.g.* lipofectamine, DEAE dextran, *etc.*
3. Physical methods which involve the use of physical methods for delivering nucleic acids into the cells *e.g.* electroporation, particle bombardment, *etc.*

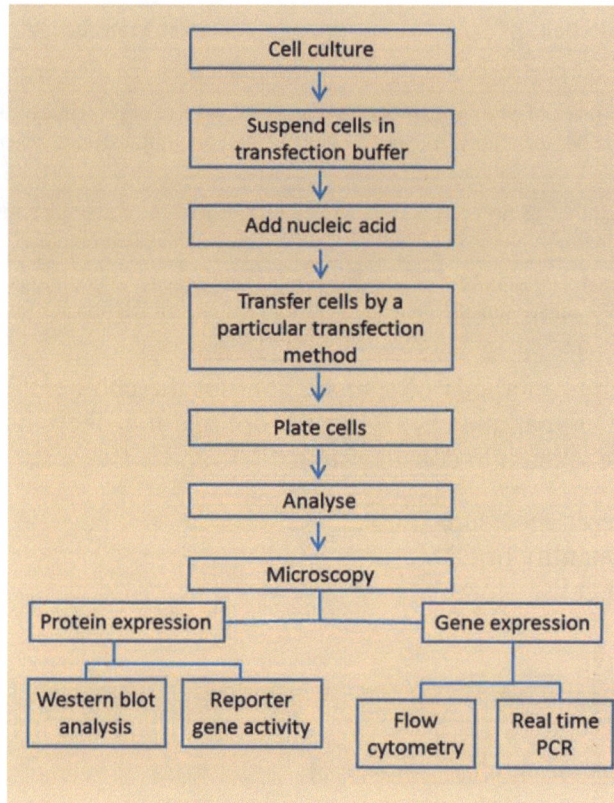

Fig. (2.1). Flow Chart Showing the General Transfection procedure.

2.1. Chemical Methods of Transfection

The chemical methods involve the use of carrier molecule to pass through the cell membrane. The negatively charged nucleic acids interact with the positively charged carrier molecules, enabling the nucleic acids to come in contact with the negatively charged membrane and thus introducing the nucleic acid into the cells by endocytosis and later releasing it into the cytoplasm.

DEAE (Diethylaminoethyl)-dextran: In 1965, this method of transfection was first done by Vaheri and Pagano. The DNA is first mixed with DEAE dextran to form a complex. The positive charge to the complex is contributed by the polymer and due to which it comes into contact with negatively charged membrane. The complex is presumably incorporated by endocytosis into the cytoplasm.

Pros:

• It is one of the simplest methods.

• Its economical.

Cons:

• The efficiency is low.
• It's not acceptable to produce stable transfected cell line.

Calcium Phosphate Method: This method of Transfection was first of all done carried out in 1973 [3]. Here, DNA is first mixed with the calcium chloride, followed by addition of buffered saline phosphate solution and incubating the mixture at room temperature. This leads to the creation of co-precipitates which then adhere to the surface of cells. The precipitates are also probably up taken by endocytosis.

Pros:

• The components used in this method are cheap and easily available.
• This method is used to generate stable cell lines.

Cons:

• This method is sensitive to slight changes in temperature, pH and buffer salt concentrations.
• The transfection efficiency is less.
• Issue of toxicity to primary cells.

Lipofection: It's one of the most commonly used chemical methods of Transfection [4]. It involves use of cationic transfection lipids comprising of a positively charged head group, a flexible linker group and two or more hydrophobic tail groups. In this method, the cationic lipid is mixed with a helper lipid (*e.g.* DOPE) and thus uni-laminar liposome vesicles are formed. The nucleic acids adsorb to these vesicles and the ionic absorption to the cellular membrane occurs and is followed by endocytosis. The helper lipids are actually the neutral lipids which allow the entrapped DNA to escape the endosomes by allowing the fusion of the liposome with the membrane.

Pros:

• It's used to transfect a wide range of cell types.
• The efficiency is good.
• The method is inexpensive.
• This method delivers DNA of all sizes and even RNA.
• It's applicable for the generation of both stable and transient cell lines.

Cons:

- The efficiency is low in primary cells.
- Its cell division dependent.
- Toxicity issues.

2.2. Physical Methods of Gene Delivery

Physical methods involve the use of physical or mechanical means for introducing nucleic acids into the cytoplasm or inside the nucleus and without the use of foreign substances like lipids [5]. The physical methods of Transfection include:

Microinjection: This is one of the sophisticated techniques for the manipulation of single cells and is also used for the generation of transgenic animals. In this technique, using a micromanipulator or microscope, a fine tipped pipette is used to insert nucleic acids into the cytoplasm or directly into the nucleus.

Pros:

- The method is highly efficient.
- Used to transfer genetic material into embryonic stem cells that are used to produce transgenic organisms.
- Frequency of stable integration is far better compared to other processes.

Cons:

- It can't be used to transfect a large number of cells.
- It requires high expertise.
- The technique is time consuming and costly.

Biolistic Particle Bombardment: This technique has been used to deliver nucleic acids not only to the cultured cells but also to the cells *in vivo*. Here, the DNA is coated on the surface of microparticles which are made up of gold and tungsten. These microparticles are then accelerated towards the cell at a high speed and thus DNA gets incorporated into the cell.

Pros:

- It's fast and simple.
- Used to transfect both dividing and non dividing cells.
- Size of the gene transfected and the number of genes transfected has no limit.

Cons:

- Due to high mortality rate, the number of cells required is very high.

Electroporation: This technique was first of all reported for gene transfer studies in mouse cells. In this technique, the cells and desired molecules are placed in conductive solution and an electric pulse at an optimised voltage is generated to create temporary pores in the cell membrane through which exogenous molecules like nucleic acids can pass into the cells. This is most commonly used for cell types such as plant protoplasts. During this technique, cells are exposed to high intensity electric field that temporarily destabilises the membrane and makes it permeable to exogenous molecules present in the surrounding media. The nucleic acids move into the cells through these pores. When the electric field is turned off, the holes in the membrane reseal enclosing the nucleic acids within [6]. The method for electroporation is carried out as follows:

- Culture the cells to be transfected in complete medium till they reach to late-log phase so that we will get reasonable number of transfectants (usually 5×10^6 cells).
- Cell harvesting is done by centrifugation at $640 \times g$ for 5 minutes.
- The pellet obtained is resuspended in ice-cold electroporation buffer which is specific for particular cell line.
- Make separate aliquots of about 0.5 ml of cell suspension into desired number of electroporation cuvettes set on ice.
- Add DNA to cell suspension in the cuvettes on ice. The DNA should be purified first by density gradients and also by phenol chloroform extraction and finally by ethanol precipitation. About 1 to 10ug of linearised DNA is used for stable transformation and 10-40ug of supercoiled DNA for transient expression.
- Mix both the cell suspension and DNA in a cuvette and incubate it for 5 minutes on ice.
- Place cuvette in the electroporation apparatus and shock one or more times at the desired voltage and capacitance settings. The number of shocks and the voltage and capacitance settings will vary depending on the cell type and should be optimized.
- After electroporation procedure is done, return cuvette containing cells and DNA to ice for 10 min.

Pros:

- It is a quick technique and easy to perform.
- It doesn't alter the biological function of the target cells.
- Highly efficient.
- Can be applied to transient and stable transfections in different types of cells.

Cons:

- Cell mortality is high.

• Low efficiency in transfecting primary cells.

Laserfection: In this method, laser light is used to temporarily permeabilize a large number of cells in a very short time. The small molecules like siRNA's, plasmids *etc.* can be efficiently injected into a wide range of cell types. When the laser induces pores in the membrane, the osmotic difference between the medium and cytosol facilitates the entry of desired substances in the medium into the cell [7].

Pros:

• High Transfection efficiency due to the ability to make pores at any location on the cell.
• Less cell manipulation required.
• Works with numerous cell types.

Cons:

• Requires an expensive laser-microscope system.

2.3. Biological Methods of Transfection

Some of the viruses have been selected as the delivery vehicles due to the ability to carry foreign genes efficiently into the host systems. It involves the generation of recombinant viruses *i.e.* viruses containing the foreign gene. These recombinant viruses are then amplified in packaging cell lines [8]. The viruses are then isolated, purified, titrated and then infected into the desired cell types. Adenovirus, retrovirus, vaccinia virus and herpes simplex virus are some of the commonly used viral gene transfer vectors.

Pros:

• The transfection efficiency is very high in primary cells.

Cons:

• The cells which do not possess viral specific receptor can't be transfected by this method.
• The amplification of recombinant viruses requires packaging cell lines.
• Time consuming and costly.
• The DNA size insert is limited.

A schematic representation of all the physical and biological methods of Transfection are shown in Fig. (**2.2**).

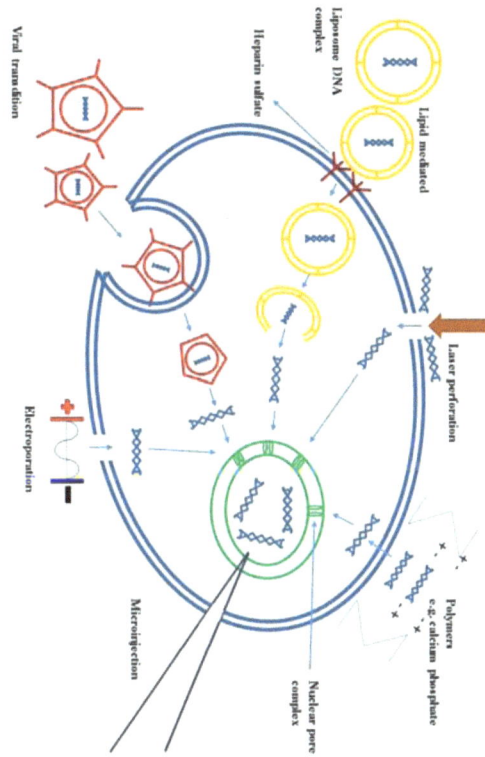

Fig. (2.2). Schematic Diagram Representing physical, chemical and biological methods of Transfection.

CONCLUDING REMARKS

Transfection is the process by which genetic material, such as DNA and double stranded RNA is inserted into mammalian cells. The insertion of DNA into a cell enables the expression, or production, of proteins using the cells own machinery and is among the widely used biotechnological approach. Here, we have discussed various techniques of DNA transfection including the biological, chemical and physical methods.

CONSENT FOR PUBLICATION

Not applicable

CONFLICT OF INTEREST

None Declare

ACKNOWLEDGEMENTS

None Declare

REFERENCES

[1]   Kim TK, Eberwine JH. Mammalian cell transfection: the present and the future. Anal Bioanal Chem 2010; 397(8): 3173-8.
      [http://dx.doi.org/10.1007/s00216-010-3821-6] [PMID: 20549496]

[2]   Kamimura K, Suda T, Zhang G, Liu D. Advances in gene delivery systems. Pharmaceut Med 2011; 25(5): 293-306.
      [http://dx.doi.org/10.1007/BF03256872] [PMID: 22200988]

[3]   Graham FL, van der Eb AJ. A new technique for the assay of infectivity of human adenovirus 5 DNA. Virology 1973; 52(2): 456-67.
      [http://dx.doi.org/10.1016/0042-6822(73)90341-3] [PMID: 4705382]

[4]   Felgner PL, Gadek TR, Holm M, *et al.* Lipofection: a highly efficient, lipid-mediated DNA-transfection procedure. Proc Natl Acad Sci USA 1987; 84(21): 7413-7.
      [http://dx.doi.org/10.1073/pnas.84.21.7413] [PMID: 2823261]

[5]   MacDonald C. Animal cell biotechnology. London: Academic Press 1992; pp. 163-5.

[6]   Huntington P, Richard H. Transfection by electroporation. Curr Protoc Mol Biol 2003; p. 3.

[7]   Rhodes K, Clark I, Zatcoff M, Eustaquio T, Hoyte KL, Koller MR. Cellular laserfection. Methods Cell Biol 2007; 82: 309-33.
      [http://dx.doi.org/10.1016/S0091-679X(06)82010-8] [PMID: 17586262]

[8]   Freshney RI. Animal cell culture: a practical approach. Washington, DC: RL Press 1986; pp. 242-71.

CHAPTER 3

Apoptosis

Abstract: Apoptosis is a critical process having widespread biological significance; regulating the differentiation, proliferation, maintenance and sculpturing organs and tissues, functioning of immune system and the elimination of defective harmful cells. The word apoptosis is derived from Greek word meaning "απόπτωσις" meaning "falling off of petals from flowers". It is actually programmed cell death essential for normal metabolism. In this chapter, we will provide general overview of various technical approaches for detecting apoptotic cells. The features, advantages and disadvantages of different methods are also discussed.

Keywords: Annexin, Caspases, DNA polymerase, Hoechst stain, Homeostasis, Luminometry, Membrane-blebbing, Necrosis, Phosphatidyl-serine, Propidium iodide, TUNEL assay.

INTRODUCTION

Apoptosis, known as programmed cell death, is highly regulated program for maintaining normal homeostasis, wound repair, developmental and defence mechanisms [1]. It is among the hot topics of biological research due to the role of apoptosis in the progression of various disorders [2]. The infecting agents and often tumour cells evade the normal induction of apoptosis as a strategy of increased survival in host. However, in some neurological disorders there are excessive levels of apoptosis, thereby, explaining the need to understand this critical phenomenon. In 1885, Fleming gave the morphological characterization of apoptosis and in 1972, John Kerr coined the term "apoptosis" [3]. Later, it was discovered that apoptosis has distinct biochemical characteristics and is a normal feature of the developmental process. Morphologically, the characteristic features of apoptotic cell include changes in refractive index of cell followed by cytoplasmic shrinkage, chromatin condensation and loss of membrane phospholipid symmetry [4]. During apoptosis, Phosphatidylserine that is normally present on the inner leaflet flips towards outer leaflet of membrane [5]. The protusions in the form of blebs or spikes appear on the cell membrane, which gives apoptotic cell a distinctive appearance. The molecular mechanisms of apoptosis involve a number of steps which ultimately result in protein cleavage, DNA degradation, membrane asymmetry and formation of apoptotic bodies from

Taseen Gul, Henah Mehraj Balkhi & Ehtishamul Haq

the breakdown of cells. Approximately, 10^{10}-10^{11} cells in the human body die every day due to apoptosis [6]. Normally, it is required for the removal of unwanted cells in the multicellular organisms, but the failure in the regulation of apoptosis can result in serious consequences. However, the events occurring during apoptosis stand in opposition to necrosis [7]. Necrosis does not include chromatin condensation, apoptotic body formation but often affects groups of adjacent cells. There is loss of membrane integrity, many internal organs lyse and inflammatory responses are provoked [8]. However, after extended incubation in *in vitro* studies, the apoptotic cells shut down metabolism and release cytoplasmic contents into the media [9]. Thus, the cells may show some of the features associated with necrosis in cell culture based studies.

3.1. Mediators of Apoptosis

Caspases constitute highly conserved protein family involved primarily in apoptosis and inflammation. In 1986, Robert Horvitz discovered that nematode worms (*C.elegans*) proceed through developmental processes without loss of cells to apoptosis. It was demonstrated that mutation in the *ced-3* gene played a key role in the process of apoptosis in these worms [10, 11]. This further led to the recognition of homologous protein family, Caspases in mammals. Caspases expressed as inactive zymogens consist of one large subunit, one small subunit and an N-terminal prodomain. Upon removal of prodomain by cleaving the zymogen at specific aspartic acid residue, the inactive zymogen gains full activity. The active caspase protein comprises of a tetramer of two heterodimers. They are the proteases with Cysteine residue at the catalytic site and cleave number of proteins for triggering the process of cell death. They cleave Focal Adhesion Kinase (FAK) which leads to the detachment of apoptotic cell from its neighbours. The disassembly of the nuclear envelope and shrinkage of nucleus occurs due to the cleavage of lamins. The cleavage of cytoskeletal proteins like tubulin, actin leads to changes in cell shape. A specific DNase known as Caspase Activated DNase (CAD) elicits DNA fragmentation through activation by Caspases [12, 13]. The Caspases are broadly categorised as initiators (Caspase 2, 8, 9, 10), executioners (Caspase 3,6,7) and inflammatory Caspases (Caspase 1,4,5) [14]. In humans, Caspase 8 and Caspase 9 initiate the apoptosis and Caspase 3 executes and commits the cell towards apoptotic fate. Another important mediator of apoptosis includes the Bcl-2 protein family. In C.elegans, it was discovered that the gene *ced-9* protects against apoptosis and loss of gene leads to increased apoptosis [15]. However, in humans Bcl-2 gene encoded protein is the homolog of *ced-9* gene product seen involved in human lymphoma [16]. Bcl-2 protein family are structurally divided into three groups; the multiregional pro-apoptotic proteins that permeabilize the mitochondrial outer membrane, the anti-apoptotic proteins that inhibit the process and the BH3 proteins that activate the pore-forming class members [17]. Bid and Bax are the pro-apoptotic members whereas Bcl-w, Bcl-2 and Bcl-XL are the anti-apoptotic members of Bcl-2 protein family. The delicate balance between pro-apoptotic and anti-apoptotic proteins inside a cell determines the ultimate fate of cell [18].

3.2. Activation of Apoptosis

The activation of apoptosis occurs due to internal as well as external stimuli. Based on the type of stimulus, the apoptotic signalling cascade is classified into extrinsic pathway and intrinsic pathway (Fig. **3.1**). In the extrinsic pathway, the apoptosis is induced by the proteins secreted from the cells of the immune system. Tumor necrosis factor alpha TNF-α is a protein secreted in response to viral infections, elevated temperatures, exposure to harmful radiations and toxic chemical agents. TNF-α turns on apoptotic process by binding to a trans membrane receptor TNFR1. This binding produces conformational changes in the cytoplasmic domain of TNF-α receptor, which leads to recruitment of a group of proteins. The ultimate protein that gets recruited is the inactive caspase-8 which gets activated due to proteolytic cleavage. The active caspase-8 known as initiator caspase, leads to the activation of downstream Caspases that trigger the self-destruction of cell. The other death receptors include the TRAIL-R1 (TNF-related apoptosis-inducing ligand receptor 1) and Fas receptor which are activated by the ligands TRAIL and Fas respectively [19, 20]. In Intrinsic pathway, the DNA damage, oxidative stress, hypoxia, high Ca^{2+} concentration trigger the apoptosis. The pathway is regulated by Bcl-2 family of proteins. During apoptosis, Bax, one of the proapoptotic members of Bcl-2 protein family, translocates from cytosol to mitochondrial membrane. This leads to increased permeability of the membrane and release of mitochondrial proteins. The most prominent protein secreted is the cytochrome c. Once released cytochrome c forms part of a protein complex known as apoptosome. The multiprotein complex consists of inactive caspase-9 which gets activated by joining the complex rather than proteolytic cleavage. The active caspase-9 activates the downstream Caspases which lead to apoptosis. The Extrinsic and Intrinsic pathways converge by activating the executional Caspases which destroy the cellular targets. Though the two pathways are activated by different stimuli but there are reports that the pathways are linked and influence each other [21, 22].

3.3. Methods for Detecting Apoptosis

Apoptosis is tightly regulated at multiple points providing opportunities to evaluate the activity of proteins involved. The measurement of apoptosis is essential to evaluate the cytotoxicity of drugs, to identify apoptosis and to elucidate the mechanism of inducing apoptosis by different compounds. Caspases are among the best targets for detecting apoptosis within cells as they become activated during apoptosis. The specific antibodies against the active form of caspase can be used for detecting their activity and thereby monitoring apoptosis. The indicators of apoptosis such as cytochrome c are also used for monitoring apoptosis. In addition, the biochemical features like DNA fragmentation and loss of membrane symmetry can also be used to determine apoptosis. The choice of a method depends upon various parameters like the type and number of cells, the pathway which induces the apoptosis and the exact method of analysis. Some of the commonly used methods for the detection of apoptosis are mentioned and briefly discussed.

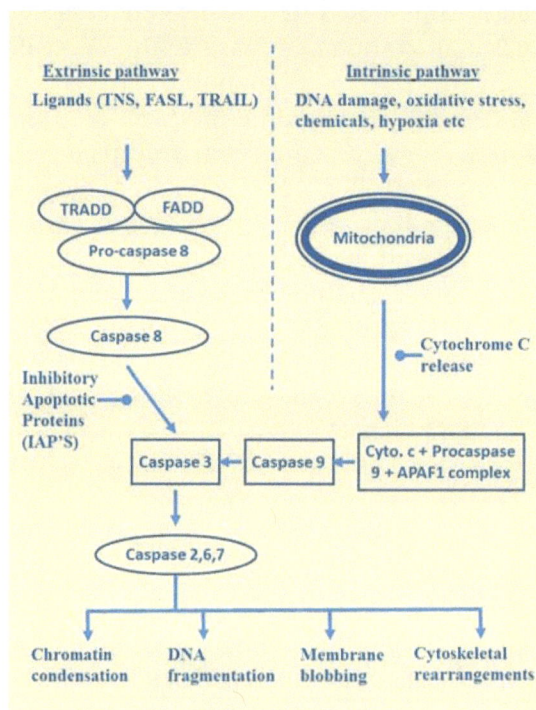

Fig. (3.1). Schematic Representation of the biochemical steps in Apoptotic pathways: The Intrinsic and Extrinsic pathways are the two major pathways that lead to the activation of apoptosis. These pathways are activated by specific stimuli and activate specific Caspase which executes the apoptotic event. The executional pathways lead to characteristic cytomorphological features including chromatin condensation, cell shrinkage formation of apoptotic bodies and finally phagocytosis of the apoptotic bodies by macrophages, neoplastic cells or adjacent parenchymal cells.

3.3.1. Analysing Cytomorphological Alterations

Cytomorphological alterations involve nuclear and cytoplasmic condensation of cells that occurs during apoptosis. For analysing these changes, cells are stained with haematoxylin and eosin and visualised by light microscopy. The single apoptotic cells can be detected by this method but it requires confirmation by other methods. However, the early events of apoptosis are not detected. Toluidine blue or methylene blue is also used for revealing intensely stained apoptotic cells by standard light microscopy. The disadvantage of the method arises due to the false positives for healthy cells having dense intracellular granules.

3.3.2. DNA Fragmentation for Detecting Apoptosis

The DNA fragmentation is a characteristic feature of apoptosis and forms the basis of a number of assays for detection of apoptosis. The chromosomes are first cleaved into large fragments followed by cleavage of DNA into 180-200bp fragments by the action of Ca^{2+} and Mg^{2+} dependent endonucleases. The cleaved DNA fragments can be visualised as a ladder on gel electrophoresis. However, the method of visualising DNA laddering is applicable only when a large number of cells are involved in apoptosis. When only a few cells are apoptotic in samples, alternate methods are used. TUNEL is one of the commonly used assays for detecting apoptosis. In 1992, Garvieli, Sherman and Bensasson demonstrated the method of *In situ* labelling of DNA breaks on a variety of tissues [23]. In TUNEL assay, the fragmented DNA ends are labelled with terminal deoxynucleo-tidyltransferase (TdT). The TdT polymerase catalyzes the addition of labelled deoxynucleotides to the 3'OH ends of DNA. The nucleotides are usually labelled with fluorescein, biotin or DIG and detected by standard immuno-fluorescent and immuno-histochemical techniques. The biotin labelled nucleotides are detected by using avidin conjugated to a reporter (*e.g.* alkaline phosphatase), the nucleotides labelled with FITC are directly detected by immuno-fluorescence and the DIG labelled nucleotides are detected by conjugated anti-DIG secondary antibody (Fig. **3.2**). The adherent cells, smears, cytopsin preparations, cryopreserved tissue sections are first fixed in paraformaldehyde solutions, washed with phosphate buffer saline and then permeabilised (Triton X-100 and sodium citrate). This is followed by the addition of TUNEL reaction mixture and finally the analysis of samples is done depending upon the type of labelling of nucleotides [24]. In 1993, Wijsman introduced a modification in TUNEL assay known as ISEL (*In situ* end-labelling technique). The principle of ISEL is similar to TUNEL except for the use of DNA Pol I to label the 3'OH DNA nicks rather than TdT. However, ISEL takes longer time to perform and is less sensitive. TUNEL assay is considered best for detection of apoptosis because it works on a number of cell types. However, the assay is critical to fixation and thus makes sample size critical. It is expensive and time consuming. Identification of apoptosis by TUNEL assay is not sufficient because the chromosomal DNA degradation also occurs in necrosis and therefore the assay is not specific. The examination of location and morphological characteristics of cells labelled by TUNEL assay are also required.

The dyes like Hoechst 33342 and DAP1 have also been used for the detection of apoptosis. These dyes become highly fluorescent upon binding to DNA and thus make the chromatin condensation of apoptotic cells readily visible. The dyes like Propidium Iodide have also been used in conjugation with other methods for the determination of apoptosis. The dye is impermeable to membrane and cannot pass the live cells, thereby stains dead cells only. Propidium Iodide intercalates between the bases and its fluorescence gets enhanced on binding to DNA. The dye is suitable for flow cytometry, fluorescence microscopy, fluorimetry *etc.* [25 - 27].

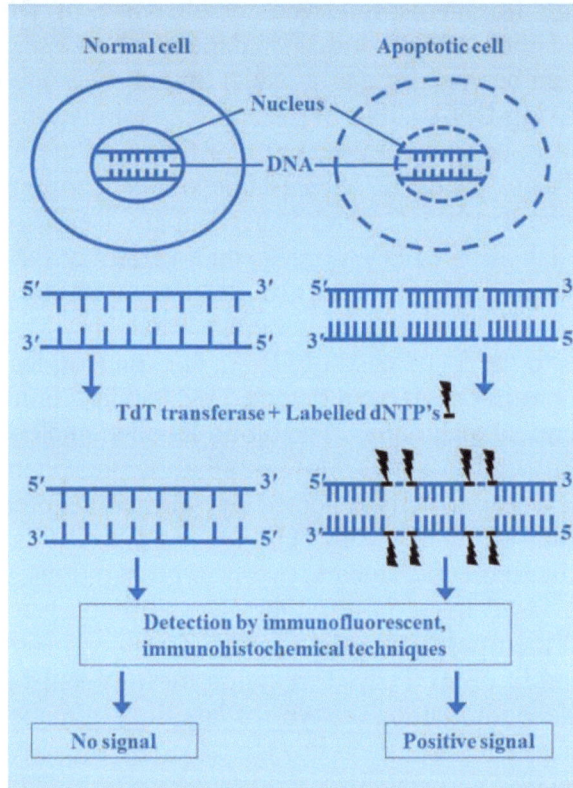

Fig. (3.2). Schematic representation of Tunnel assay: In this assay, labeled deoxynucleotides (dNTPs) are added to fragment DNA by the enzyme TdTtransferase. The DNA in live cells is intact so the enzyme is unable to add labelled deoxynucleotides to the intact DNA. However, in apoptotic cells double or single strand breaks occur in the genomic DNA and the labelled dNTPs are incorporated, thereby giving a positive signal.

3.3.3. Mitochondrial Markers for Detecting Apoptosis

The disruption of active mitochondria is a characteristic feature of early stages of apoptosis. It includes alterations in the membrane potential and oxidation-reduction potential of mitochondria. These changes are presumed to occur due to change in mitochondrial membrane permeability allowing passage of molecules across it. The common method for detecting the change in mitochondrial membrane permeability involves a fluorescent lipophilic cationic dye 5,5′,6,6′-tetrachloro-1,1′,3,3′-tetraethylbenzimidazolcarbocyanine iodide (JC-1). The dye has been used as an indicator of mitochondrial membrane permeability in a variety of cell types as well as in tissues and isolated mitochondria [28]. The method was first studied in K562 and U937 cell lines. In monomeric form, JC-1 emits at 527nm whereas the aggregated form is associated with a significant shift in emission (590nm).The dye accumulates inside mitochondria in non-apoptotic cells, remain as aggregates, and thereby emit bright red fluorescence. However, in apoptotic cells, the mitochondrial membrane does not maintain electrochemical gradient and leads to diffusion of dye into the cytoplasm and therefore green fluorescence specific for monomeric forms is emitted (Fig. **3.3**). JC-1 staining can be used for both suspension as well as adherent cells. The cells are first seeded depending upon the recommended density for a particular cell type. For suspension cells, the staining solution (JC-1 stain dissolved in DMSO) and growth media are mixed and then the cells are incubated at 37°C in a humidified atmosphere containing 5% CO_2. The cells are then centrifuged and pellet is resuspended in staining buffer. The stained cells are then observed by fluorescence microscopy or by flow cytometric assay and measured by fluorimetric assay. For adherent cells, firstly the media is aspirated out and then the cells are incubated with staining solution. Then, the staining solution is aspirated out, cells are washed with buffer or growth media and visualised under fluorescence microscope. Since, the membrane potential is sensitive to changes in pH and temperature; therefore the reagents should be carefully checked. The dye is sensitive to light and thus incubations need to be done in dark and the samples should be analysed promptly after completion of staining. Apart from JC-1, other dyes such as rhodamine-123 and DiOC6 have also been used to estimate the apoptotic changes but they are not reliable due to less sensitivity and specificity.

One of the critical players of apoptosis which has been largely used for the detection of apoptosis is the Cytochrome c. It is a water soluble protein located in the intermembrane space of mitochondria. The stimulation of apoptosis leads to the release of cytochrome c from mitochondria into the cytosol, where it forms a complex with Apaf-1 and activates Caspase 9. This leads to the activation of Caspase 3 and the downstream executioner Caspases. The translocation of cytochrome c from mitochondria towards cytosol paves way for the detection of apoptosis. The mitochondrial and cytosolic fractions are isolated and western blotting technique using cytochrome c antibody is used to study its translocation and thereby the event of apoptosis [29, 30]. Immunohistochemistry and ELISA can also be done for the detection of cytochrome c. BCl-2 protein family which have a critical role in apoptosis are also detected by western blotting and immunohistochemistry.

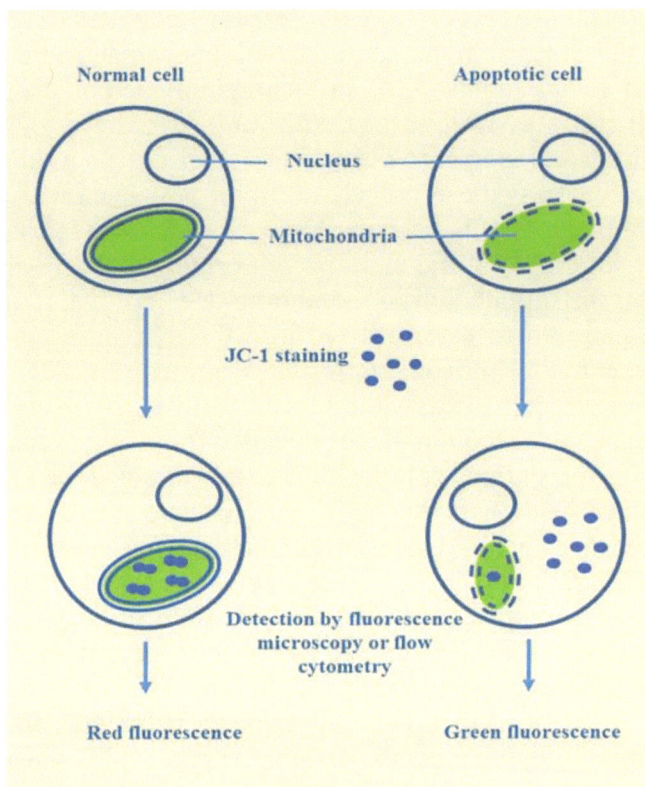

Fig. (3.3). Diagrammatic representation of JCI staining: In live cells the JCI stain permeates the cell membrane and accumulates in the mitochondria where it oligomerizes to emit red florescence, however in apoptotic cells, the mitochondrial membrane integrity is lost, the JCI can't form oligomers and remains scattered in the cytosol to emit green florescence.

For the assessment of integrity of outer mitochondrial membrane, cytochrome c oxidase assay is carried out. Cytochrome c oxidase is an enzyme involved in coupling of electron transport with oxidative phosphorylation. It is located on the inner mitochondrial membrane and the assay is based on the decrease in absorbance of ferrocytochrome to ferricytochrome due to its oxidation caused by cytochrome c oxidase. The assay is carried out in presence and absence of detergent (n-dodecyl β-D-maltoside). The detergent at low concentrations maintains the cytochrome c oxidase dimer in solution. The ratio between the cytochrome c oxidase activity in presence and absence of detergent is a measure of integrity of outer mitochondrial membrane, since the membrane is the barrier for the entry of cytochrome c into organelle. The assay requires the use of freshly prepared samples as the frozen tissues may cause rupture of sub cellular organelles.

3.3.4. Measuring Changes in Cell Membrane to Detect Apoptosis

The plasma membrane of viable cells possess asymmetric distribution of phospholipids was first observed in erythrocytes and later in nucleated cell types. During apoptosis, the asymmetry of phospholipids on cell membrane is lost due to the flipping of Phosphatidylserine towards outer leaflet of membrane [31]. Detecting the change in phospholipid asymmetry is one of the ways to detect apoptosis. Annexin-V is a phospholipid binding protein that has a high affinity for Phosphatidylserine and conjugating Annexin V to a fluorescent molecule or dye can be used to label apoptotic cells. Koopman *et al.* first described a method of using Annexin V labelled to applied hapten (*i.e.* FITC/Biotin) to detect apoptosis. The labelled Annexin V binds to the Phosphatidylserine residues in presence of Ca^{2+}. Annexin V cannot penetrate the phospholipid bilayer and as such does not bind normal live cells (Fig. **3.4**). In dead cells, the integrity of plasma membrane is lost and the labelled Annexin V binds to inner leaflet of membrane. However, to discriminate between apoptotic and dead cells, DNA stain such as propidium iodide can be added. Thus the apoptotic, viable and dead cells can be distinguished on the basis of the double labelling for Propidium Iodide and Annexin V and subsequently analysed by flow cytometry and fluorescence microscopy. However, for quantification of apoptotic cells, flow cytometry can be used by using a cell suspension prepared from cells or tissues. The assay has also been used for detecting the apoptotic cells *in situ*. The biotin labelled Annexin is injected into mice and then followed by dissection and formalin fixing of tissues. The tissue sections are incubated with streptavidin conjugated peroxidase and finally visualised. Thus, the apoptosis could be detected even in developing mouse embryos by this technique. The phosphatidyl serine exposure is a universal phenomenon that occurs during early apoptosis prior to detection of DNA strand fragmentation, providing advantage for detection of apoptosis by Annexin V assay. The assay is fast, simple, sensitive and independent of species or apoptosis inducing systems. However, the assay cannot be used for the cells which express high levels of Phosphatidylserine on their outer membrane [32 - 35].

3.3.5. Measuring Caspase Activity for the Detection of Apoptosis

Caspases are the central mediators of proteolytic cascade leading to apoptosis of cells. The activation of Caspases is tightly regulated by transcription and by anti-apoptotic polypeptides and therefore, the detection of Caspase activity is among the best methods for detecting apoptosis [36]. Analysing procaspase processing by Immunoblotting, analysing enzyme activity by cleaving synthetic substrates and the Immunoblotting analysis of cleavage of Caspase substrates are among the different methods for the determination of Caspase activation. One of the important methods for analysing the activation of Caspases involves the detection of their target molecule Poly ADP-ribose polymerase (PARP) by Immunoblotting [37]. PARP is a nuclear enzyme involved in DNA repair and is cleaved by Caspase 3 during apoptosis.

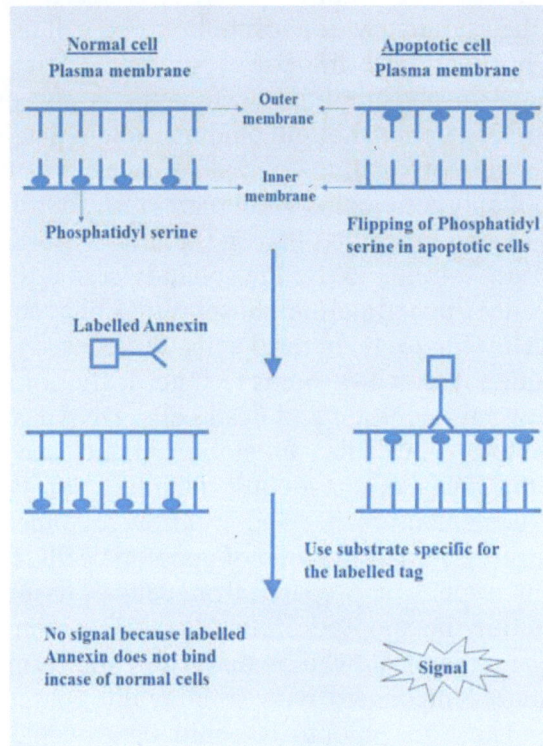

Fig. (3.4). Diagrammatic representation of Annexin V assay: During apoptosis, the phosphatidyl serine (●) which is normally present on the inner leaflet of membrane flips towards the outer leaflet in apoptotic cells. The flipping of phosphatidyl serine is detected by the labelled Annexin V protein. The labelled annexin binds to apoptotic cell and gets detected whereas normal cells do not.

The Caspase activity assays are among the commonly used methods for the detection of apoptosis and involve the use of synthetic substrates. The technique was first described by Pennington and Thornberry. The tetra peptide sequence mimicking the specific cleavage sites of Caspases are synthesised and conjugated to a detectable entity (chromophore or fluorophore).The Caspase activity assays are colorimetric, luminometric or fluorimetric depending upon the type of tetra peptide substrate used. For luminometric Caspase activity assays, the substrates used are Z-LETD-aminoluceferin (Caspase 8 substrate), Z-DEVD-aminolucefrin (Caspase 3/7 substrate), Z-LETD-aminoluciferin (Caspase 9 substrate) and Z-VEID-aminoluciferin (Caspase 6 substrate). The enzymes used are the luciferases which catalyze the reaction and luminescence is emitted. The buffers are optimised for monitoring specific Caspase activity. When the Caspases are inactive, the substrates are not cleaved and thus no light is produced. However, upon cleavage of substrate by active Caspase, light is generated which is proportional to the Caspase activity. For colorimetric detection of Caspase activity, the substrates used are labelled with chromophore p-nitroaniline. The chromogen is released out upon cleavage by Caspase and produces yellow colour which is then measured spectrophotometrically at 405nm. The colorimetric detection of Caspase 3 activity provides quantitative measurement of protease activity which is a regulatory event in the programmed cell death process. The substrates like profluorescent DEVD peptide-rhodamine 110 are used for the fluorescent detection of Caspase 3/7 activity (Fig. **3.5**). The substrates mixed with buffers is added directly to the culture dish and incubated. The cells are permeabilised so that they release the Caspase and the fluorescent product proportional to the Caspase activity in the sample gets accumulated [38 - 40]. The enzyme activity assays are sensitive to numbers of factors, therefore, optimal pH, salt concentration and optimal buffer composition should be maintained for carrying out the assay.

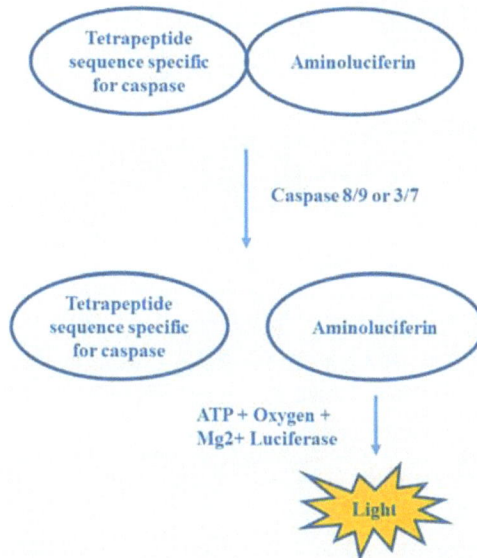

Fig. (3.5). Diagrammatic representation of Caspase activity assay: Caspase are activated during apoptosis and thus measuring Caspase activity gives a direct proportion of apoptotic activity inside a cell. Caspase act on specific substrates and conjugating these substrates to chromophores or flurophores are used for the detection and quantification of Caspase activity and thereby apoptosis.

CONCLUDING REMARKS

The detection of apoptosis in cell culture studies is an important cytochemical technique for medical studies. The commonly used methods developed for the detection of apoptosis have been derived from the properties associated with the living cell. However, the assay methods based on the highly specific biochemical and molecular mechanisms of apoptosis have also been developed. The apoptotic cell after extended incubation loses membrane integrity and release cellular contents into the culture medium and resembles necrosis. Thus, knowing the kinetics of cell death process in the system under consideration is critical for differentiating the two events. Using more than one method for the detection of apoptosis is best for confirming the event. One of the methods should detect an early apoptotic event and the other an executional event. However, multiplexing more than one assay can provide better insight of the process and eliminate the need to repeat the work.

CONSENT FOR PUBLICATION

Not applicable

CONFLICT OF INTEREST

None Declare

ACKNOWLEDGEMENTS

None Declare

REFERENCES

[1]   Elmore S. Apoptosis: a review of programmed cell death. Toxicol Pathol 2007; 35(4): 495-516.
        [http://dx.doi.org/10.1080/01926230701320337] [PMID: 17562483]

[2]   Chin S, Hughes MP, Coley HM, Labeed FH. Rapid assessment of early biophysical changes in K562
        cells during apoptosis determined using dielectrophoresis. Int J Nanomed 2006; 1(3): 333-7.
        [PMID: 17717973]

[3]   Kerr JF, Wyllie AH, Currie AR. Apoptosis: a basic biological phenomenon with wide-ranging
        implications in tissue kinetics. Br J Cancer 1972; 26(4): 239-57.
        [http://dx.doi.org/10.1038/bjc.1972.33] [PMID: 4561027]

[4]   Hengartner MO. Programmed cell death.C elegans. New York: Cold Spring Harbour Laboratory Press
        1997; pp. 383-496.

[5]   Williamson P, van den Eijnde S, Schlegel RA. Phosphatidylserine exposure and phagocytosis of
        apoptotic cells. Methods Cell Biol 2001; 66: 339-64.
        [http://dx.doi.org/10.1016/S0091-679X(01)66016-3] [PMID: 11396011]

[6]   Hirsch T, Marchetti P, Susin SA, *et al*. The apoptosis-necrosis paradox. Apoptogenic proteases
        activated after mitochondrial permeability transition determine the mode of cell death. Oncogene
        1997; 15(13): 1573-81.

[http://dx.doi.org/10.1038/sj.onc.1201324] [PMID: 9380409]

[7]   Leist M, Jäättelä M. Four deaths and a funeral: from caspases to alternative mechanisms. Nat Rev Mol Cell Biol 2001; 2(8): 589-98.
[http://dx.doi.org/10.1038/35085008] [PMID: 11483992]

[8]   Riss TL, Moravec RA. Use of multiple assay endpoints to investigate the effects of incubation time, dose of toxin, and plating density in cell-based cytotoxicity assays. Assay Drug Dev Technol 2004; 2(1): 51-62.
[http://dx.doi.org/10.1089/154065804322966315] [PMID: 15090210]

[9]   Hedgecock EM, Sulston JE, Thomson JN. Mutations affecting programmed cell deaths in the nematode Caenorhabditis elegans. Science 1983; 220(4603): 1277-9.
[http://dx.doi.org/10.1126/science.6857247] [PMID: 6857247]

[10]   Ellis HM, Horvitz HR. Genetic control of programmed cell death in the nematode C. elegans. Cell 1986; 44(6): 817-29.
[http://dx.doi.org/10.1016/0092-8674(86)90004-8] [PMID: 3955651]

[11]   Saraste A, Pulkki K. Morphologic and biochemical hallmarks of apoptosis. Cardiovasc Res 2000; 45(3): 528-37.
[http://dx.doi.org/10.1016/S0008-6363(99)00384-3] [PMID: 10728374]

[12]   Gerald K. Cell and molecular biology: concepts and experiments. USA: John and Wiley Sons 2013; pp. 330-40.

[13]   Cohen GM. Caspases: the executioners of apoptosis. Biochem J 1997; 326(Pt 1): 1-16.
[http://dx.doi.org/10.1042/bj3260001] [PMID: 9337844]

[14]   Hengartner MO, Ellis RE, Horvitz HR. C.elegans cell survival gene ced-9 protects cells from programmed cell death. Nature 1992; 356: 494-9.
[http://dx.doi.org/10.1038/356494a0] [PMID: 1560823]

[15]   Tsujimoto Y, Croce CM. Analysis of the structure, transcripts, and protein products of bcl-2, the gene involved in human follicular lymphoma. Proc Natl Acad Sci USA 1986; 83(14): 5214-8.
[http://dx.doi.org/10.1073/pnas.83.14.5214] [PMID: 3523487]

[16]   Shamas A, Kale J, Leber B, Andrews DW. Mechanism of action of Bcl-2 family proteins. Cold Spring Harb Perspect Biol 2013; 5(4): 008714.
[http://dx.doi.org/10.1101/cshperspect.a008714]

[17]   Daniel PT, Schulze-Osthoff K, Belka C, Güner D. Guardians of cell death: the Bcl-2 family proteins. Essays Biochem 2003; 39: 73-88.
[http://dx.doi.org/10.1042/bse0390073] [PMID: 14585075]

[18]   Wajant H. Death receptors. Essays Biochem 2003; 39: 53-71.
[http://dx.doi.org/10.1042/bse0390053] [PMID: 14585074]

[19]   Hengartner MO. The biochemistry of apoptosis. Nature 2000; 407(6805): 770-6.
[http://dx.doi.org/10.1038/35037710] [PMID: 11048727]

[20]   Parone P, Priault M, James D, Nothwehr SF, Martinou JC. Apoptosis: bombarding the mitochondria. Essays Biochem 2003; 39: 41-51.
[http://dx.doi.org/10.1042/bse0390041] [PMID: 14585073]

[21]   Zou H, Henzel WJ, Liu X, Lutschg A, Wang X. Apaf-1, a human protein homologous to C. elegans CED-4, participates in cytochrome c-dependent activation of caspase-3. Cell 1997; 90(3): 405-13.
[http://dx.doi.org/10.1016/S0092-8674(00)80501-2] [PMID: 9267021]

[22]   Igney FH, Krammer PH. Death and anti-death: tumour resistance to apoptosis. Nat Rev Cancer 2002; 2(4): 277-88.
[http://dx.doi.org/10.1038/nrc776] [PMID: 12001989]

[23]   Gavrieli Y, Sherman Y, Ben-Sasson SA. Identification of programmed cell death *in situ via* specific labeling of nuclear DNA fragmentation. J Cell Biol 1992; 119(3): 493-501.

[http://dx.doi.org/10.1083/jcb.119.3.493] [PMID: 1400587]

[24] Grasl-Kraupp B, Ruttkay-Nedecky B, Koudelka H, Bukowska K, Bursch W, Schulte-Hermann R. *In situ* detection of fragmented DNA (TUNEL assay) fails to discriminate among apoptosis, necrosis, and autolytic cell death: a cautionary note. Hepatology 1995; 21(5): 1465-8.
[PMID: 7737654]

[25] Crissman HA, Wilder ME, Tobey RA. Flow cytometric localization within the cell cycle and isolation of viable cells following exposure to cytotoxic agents. Cancer Res 1988; 48(20): 5742-6.
[PMID: 2458829]

[26] Kapuscinski J. DAPI: a DNA-specific fluorescent probe. Biotech Histochem 1995; 70(5): 220-33.
[http://dx.doi.org/10.3109/10520299509108199] [PMID: 8580206]

[27] Zink D, Sadoni N, Stelzer E. Visualizing chromatin and chromosomes in living cells. Methods 2003; 29(1): 42-50.
[http://dx.doi.org/10.1016/S1046-2023(02)00289-X] [PMID: 12543070]

[28] Cossarizza. A new method for the cytofluorimetric analysis of mitochondrial membrane potential using the J-aggregate forming lipophilic cation 5,5′,6,6′-tetrachloro-1,1′,3,3′-tetraethylben-zimidazolcarbocyanine iodide (JC-1). Biochem Biophys Res Commun 1993; 19: 11-30.

[29] Li P, Nijhawan D, Budihardjo I, *et al.* Cytochrome c and dATP-dependent formation of Apaf-1/caspase-9 complex initiates an apoptotic protease cascade. Cell 1997; 91(4): 479-89.
[http://dx.doi.org/10.1016/S0092-8674(00)80434-1] [PMID: 9390557]

[30] Scorrano L, Ashiya M, Buttle K, *et al.* A distinct pathway remodels mitochondrial cristae and mobilizes cytochrome c during apoptosis. Dev Cell 2002; 2(1): 55-67.
[http://dx.doi.org/10.1016/S1534-5807(01)00116-2] [PMID: 11782314]

[31] Zachowski A. Phospholipids in animal eukaryotic membranes: transverse asymmetry and movement. Biochem J 1993; 294(Pt 1): 1-14.
[http://dx.doi.org/10.1042/bj2940001] [PMID: 8363559]

[32] Koopman G, Reutelingsperger CP, Kuijten GA, Keehnen RM, Pals ST, van Oers MH. Annexin V for flow cytometric detection of phosphatidylserine expression on B cells undergoing apoptosis. Blood 1994; 84(5): 1415-20.
[PMID: 8068938]

[33] van Engeland M, Nieland LJ, Ramaekers FC, Schutte B, Reutelingsperger CP. Annexin V-affinity assay: a review on an apoptosis detection system based on phosphatidylserine exposure. Cytometry 1998; 31(1): 1-9.
[http://dx.doi.org/10.1002/(SICI)1097-0320(19980101)31:1<1::AID-CYTO1>3.0.CO;2-R] [PMID: 9450519]

[34] Van den Eijnde SM, Boshart L, Reutelingsperger CP, De Zeeuw CI, Vermeij-Keers C. Phosphatidylserine plasma membrane asymmetry *in vivo:* a pancellular phenomenon which alters during apoptosis. Cell Death Differ 1997; 4(4): 311-6.
[http://dx.doi.org/10.1038/sj.cdd.4400241] [PMID: 16465246]

[35] O'Brien IE, Reutelingsperger CP, Holdaway KM. Annexin-V and TUNEL use in monitoring the progression of apoptosis in plants. Cytometry 1997; 29(1): 28-33.
[http://dx.doi.org/10.1002/(SICI)1097-0320(19970901)29:1<28::AID-CYTO2>3.0.CO;2-9] [PMID: 9298808]

[36] Earnshaw WC, Martins LM, Kaufmann SH. Mammalian caspases: structure, activation, substrates, and functions during apoptosis. Annu Rev Biochem 1999; 68: 383-424.
[http://dx.doi.org/10.1146/annurev.biochem.68.1.383] [PMID: 10872455]

[37] Lazebnik YA, Kaufmann SH, Desnoyers S, Poirier GG, Earnshaw WC. Cleavage of poly(ADP-ribose) polymerase by a proteinase with properties like ICE. Nature 1994; 371(6495): 346-7.
[http://dx.doi.org/10.1038/371346a0] [PMID: 8090205]

[38]   Thornberry NA, Lazebnik Y. Caspases: enemies within. Science 1998; 281(5381): 1312-6.
        [http://dx.doi.org/10.1126/science.281.5381.1312] [PMID: 9721091]

[39]   Pennington MW, Thornberry NA. Synthesis of a fluorogenic interleukin-1 beta converting enzyme
        substrate based on resonance energy transfer. Pept Res 1994; 7(2): 72-6.
        [PMID: 8012123]

[40]   Green DR. Apoptotic pathways: paper wraps stone blunts scissors. Cell 2000; 102(1): 1-4.
        [http://dx.doi.org/10.1016/S0092-8674(00)00003-9] [PMID: 10929706]

CHAPTER 4

Cell Cytotoxicity, Viability and Proliferation

Abstract: Cell viability and cytotoxicity assays are used usually for screening of drugs and cytotoxicity tests of chemicals. Cell viability, cell proliferation and many important live-cell functions can be stimulated or monitored with various chemical and biological reagents. The present chapter is aimed to discuss the various available methods (staining and non-staining) for the assessment of cell viability and proliferation.

Keywords: Biolumniscence, Bromodeoxyuridine and Clonogenic assay, Carboxyfluorescein diacetatesuccinimidyl ester, Hemocytometer, LDH assay, MTT assay, Trypan blue.

INTRODUCTION

It is essential for a number of biological experiments to measure number of surviving cells, proliferating cells or dead cells to evaluate the response of cells in culture after treatment with various external stimuli. For such an evaluation, appropriate choice of a method depends on the specific type and number of cells used, expected outcome as well as the cellular process under evaluation. The aim can be fulfilled by several methods, *e.g.*, by counting cells that either includes or exclude a stain/dye, monitoring the release of proteins after cell lysis, and measuring the incorporation of radioactive nucleotides during cell proliferation. The cell killing ability of a chemical compound or a mediator refers to the cell cytotoxicity. It is independent of the mechanism of cell death. The number of healthy cells, whether dividing or quiescent refers to the cell viability. However, the measurement of actively dividing cells in a sample refers to the cell proliferation. The quick and accurate measurement of cell viability or cell proliferation is an important requirement in *in vitro* and *in vivo* experimental situations. They are used for screening of drugs, serum batch testing, and growth factor activity and to assess the cytotoxic, mutagenic or carcinogenic effects of a chemical compound. These assays are based on cellular functions like cell membrane permeability, enzyme activity, cell adherence, ATP production, colony formation, nucleotide uptake *etc.*

Taseen Gul, Henah Mehraj Balkhi & Ehtishamul Haq

4.1. Trypan Blue Assay

In 1904, Trypan blue was synthesised for the first time by Paul Ehrlich from Toluidine. It is a vital stain also known as Niagara Blue or Diamine blue which gives characteristic blue colour selectively to dead tissues or cells [1].

The mechanism of assay is based on the chemical property of Trypan blue stain. The viable cells have intact plasma membranes and the stain due to its negative charge does not interact and pass through the cell membrane. However, if the membrane is damaged, it easily enters inside the cells and thus the cells with damaged membranes take up the stain readily and retain the blue colour which could be easily observed under a microscope. The method excludes the live cells from staining and as such it is also described as a dye exclusion method. Thus the staining procedure allows us to discriminate between viable cells and damaged or dead cells.

Assay is used for screening of various compounds for different activities such as anti-proliferation, cytotoxic compounds *etc*. For carrying out trypan blue assay, the samples under consideration are first treated with a test compounds and then the residual cell viability in the attached and detached cell populations is determined by Trypan blue. Usually 0.5% solution of Trypan blue in PBS is used. It is filtered first so as to get rid of particles which would disturb the counting process. After complete treatment of cells with different compounds cells are suspended in complete media and an aliquot of cell suspension being tested for viability is pelleted by centrifugation. Pellet is resuspended in PBS so as to achieve optimal concentration of cells for counting purpose. Next Trypan blue dye (0.5% Trypan Blue in 1:5 ratio) is added to the preparation and a small amount of Trypan Blue-cell suspension mixture is pipetted into both chambers of the hemocytometer carefully Hemocytometer chambers should not be pressure filled or under filled. Count the cells starting with chamber 1 of hemocytometer and include all the cells in the 1 mm centre square and four 1 mm corner squares Count of viable and non-viable cells is kept separate. While counting the cells which are present on top and left touching middle line of the perimeter of each square are counted while as cells touching the middle line at bottom and right sides are not counted. If cells are clustered then the entire procedure is repeated to disperse them and to reach an appropriate dilution factor. Counting procedure is repeated many times to ensure accuracy. A single square of the hemocytometer represents a total volume of 0.1 mm^3 or 10^{-4} cm^3. Since 1 cm^3 is equivalent to approximately 1 ml, the subsequent cell concentration per ml (and the total number of cells) is determined as:

$$\text{Cells per ml} = \text{Average count per square} \times \text{Dilution factor} \times 10^4$$

Percentage of viable cells is calculated as follows:

$$\text{Viable cells (\%)} = \frac{\text{Total number of viable cells per ml of aliquot} \times 5 \times 100}{\text{Total number of cells per ml of aliquot}}$$

5= the dilution factor for Trypan blue

The method is rapid, inexpensive and requires small fraction of cells from a cell population. It is usually used to determine the cell number per ml in batch cultures.

The disadvantage of Trypan blue dye exclusion assay is that the viability of cells is elucidated indirectly from cell membrane integrity. Thus there is a possibility that cell's viability may have been compromised even though its membrane integrity is maintained. Conversely, cell may have temporarily lost its integrity but may be able to repair itself and become fully viable. Further, the assay cannot distinguish between necrotic and apoptotic cells.

4.2. Neutral Red (3-Amino-7-Dimethyamino-2-Methylphenazine Hydrochloride)

The assay based on neutral red was first of all described by Borenfreund and Puerner (1984) [2]. The assay system measures living cells through the uptake of neutral red. The uptake of dye occurs through active transport and then the dye gets accumulated into the lysosomes of live cells. The cells are then washed and fixed. The dye is then liberated by acidified ethanol solution. The degree of cytotoxicity is determined by the amount of dye-uptake by the cells in the culture. Quantification is done spectrophotometrically at 540nm. However, the neutral red dye is rarely used in high density formats due to multiple addition, washes and dye extraction step [3].

4.3. Calcein AM

Calcein O, O'-diacetatetetrakis (acetoxymethyl) ester is a non fluorescent hydrophobic cell permeable compound which becomes fluorescent on hydrolysis [4]. On addition of Calcein AM ester to cells in culture, it diffuses into the cells passes the plasma membrane of cells and reacts with unspecific cytosolic esterases. The reaction product of Calcein AM ester is Calcein (a hydrophilic fluorescent compound (Fig. **4.1**). It does not penetrate membranes because Calcein is retained within the cells. Calcein chelates with low-mass labile iron as a result of which fluorescence is quenched, and the degree of quenching gives an estimate of the amounts of chelatable iron. Calcein AM is used for labeling live cells which can be detected and further analyzed either by flow cytometry or fluorescence imaging. The increased hydrophobicity of the acetomethoxy(AM) derivative of Calcein allows this particular dye to readily enter viable cells. The dead cells lack esterase activity and as such only viable cells are labelled. The quantification is done using a fluorescent plate reader with excitation and emission wavelengths of 470nm and 520nm respectively.

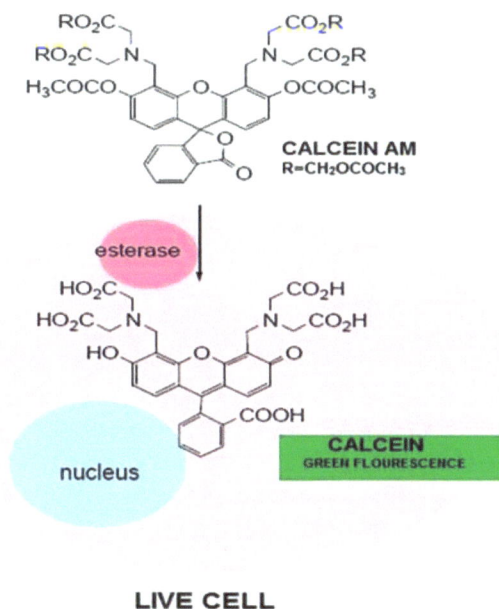

Fig. (4.1). Basic Principle of Calcein AM assay. Calcein AM is rapidly taken up by live cells and is converted to free, fluorescent Calcein by intracellular esterases.

The calcein-acetoxymethyl ester method is a commonly used method to assay the intracellular labile iron pool localized in the cytosol and not in a membrane-limited compartment. Most of the cellular, low-mass iron is temporarily localized in the lysosomal compartment as a consequence of the autophagic degradation of ferruginous materials, such as mitochondrial complexes and ferritin. This content cannot be quantified by the Calcein-AM method as Calcein does not traverse lysosomal membranes and it does not sequester iron at the low lysosomal pH. This is one of the main drawbacks of this assay. However, the assay is rapid and has better retention than other fluorescent compounds. The assay is suitable for proliferating and non-proliferating cells as well as adherent and suspension cultures. It is useful in a wide variety of studies like cell viability, apoptosis, cytotoxicity, chemotaxis and multidrug resistance *etc.* [5].

4.4. CFSE

Carboxyfluorescein diacetatesuccinimidyl ester (CFSE) is a colourless, non-fluorescent compound that diffuses passively into the cells. It's degraded by intracellular esterases to produce fluorescent carboxyfluorescein succinimidyl ester. The ester groups react with intracellular amines resulting in the formation of conjugates that are fluorescent and very well retained. The concentration of the dye should be kept as low as possible so as to reduce potential artefacts from overloading. About 0.5-5uM CFSE is needed for viability assays whereas upto 25uM are required for microscopy applications. Cells labelled with CFSE can be visualised by fluorescent microscopy using standard fluorescein filter sets or analysed by flow cytometry with a 488nm excitation source. CFSE is the best option for assessing number of cellular divisions that a population has undergone [6].

4.5. Sulphorhodamine B Assay

It's usually carried out to indicate the degree of toxicity caused by the test material. It's a means of evaluating overall biomass by staining of cellular proteins. The cells are first washed, fixed and stained with the dye. The changes in total biomass are detected by monitoring the amount of dye incorporated by cells. The dye is then extracted from the cells by using tris base buffer solution. Quantification is done spectrophotometrically by measuring absorbance at 565nm [7].

4.6. Assays Based on Metabolic Activity

Metabolic activity is an indicator of cell viability and proliferation. The metabolic activity assays are widely used to measure cell proliferation and cytotoxicity of a cell population in response to external factors. The colorimetric assays are based on the principle that the live cells due to their active enzymatic machinery have the ability to utilise a colourless substrate and convert it into a detectable coloured product. The MTT assay is one of the colorimetric assays used for measuring the cell proliferation rate [8, 9].

MTT, 3(4, 5 dimethylthiazol 2 yl) 2, 5 diphenyltetrazoliumbromide is a tetrazolium salt and appears yellowish when dissolved in a suitable solvent. MTT is reduced to generate insoluble formazan crystals and reducing equivalents by mitochondrial dehydrogenase enzyme inside metabolically active cells. Water insoluble purple coloured formazan crystals are dissolved in an acidified solvent, once solubilized it is quantified spectrophotometrically. This reduction reaction catalyzed by oxidoreductase enzymes is dependent on NAD(P)H therefore, reduction of MTT and other tetrazolium dyes depends on the cellular metabolic activity due to NAD(P)H flux.

The effect of defined conditions on proliferation of cells is quantified using the MTT colorimetric assay. Cells are cultured in triplicates for each condition in cell culture plates and allowed to grow. After treated as per the experiment, plates are incubated at 37°C and 5% CO_2. MTT is dissolved in PBS at 5 mg/ml and filtered to sterilize and remove any amount of insoluble residues. Then MTT stock solution (5 mg/ml) is added to each culture vessel or plate under investigation and incubated for 3 to 4 hr. The reaction is stopped by adding MTT solvent (0.1% NP-40 and 4mM HCI in isopropanol) to each well for dissolving the formazan crystals. This is followed by incubation period at 37°C for 20 minutes in dark. Once the crystals are dissolved, the plates are read quickly on ELISA plate reader at a test wavelength and reference wavelength of 560nm and 650nm respectively.

Inside cells, MTT is cleaved by succinate tetrazolium reductase (mitochondrial enzyme) into a coloured insoluble formazan product (Fig. **4.2**). The insoluble formazan product is then solubilised by organic solvent and then rapidly quantitated in an ELISA plate reader. The inference of higher absorbance values compared to control is an increase in the rate of cell proliferation whereas lower values indicate reduction in cell proliferation. To understand the experimental results more accurately, graphs are plotted which give a linear relationship between the absorbance values and proliferation rates under different time periods and concentrations *etc*.

The MTT assay is also carried out to monitor the development of clones and hybridomas. By converting the dye, they can be easily visualised and their growth can be detected without magnification.

Some modified tetrazolium salts are available that get converted directly into water soluble formazan by viable cells and require one step less than MTT assay.

XTT: XTT {2, 3 bis (2 methoxy 4 nitro 5 sulfophenyl) 2 H tetrazolium 5 carboxanilide} is a tetrazolium salt with higher dynamic range and higher sensitivity as compared to MTT. It is water soluble and as such the solubilisation step is not required.

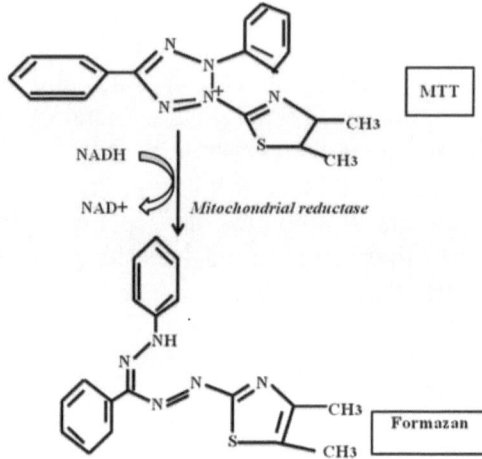

Fig. (4.2). Molecular structure of MTT and its reaction product.

MTS: MTS {3 (4, 5 dimethylthiazol 2 yl) 5 (3 carboxymethoxyphenyl) 2 (4 sulfophenyl) 2 H tetrazolium} yields a formazan product when treated with phenazine methosulfate. The liberated formazan product has an absorbance maximum at 490 – 500 nm. MTS assay offers the convenience of directly adding the reagent to the cell culture without the intermittent steps required as in the MTT assay. However, due to skipping of these steps, MTS assay is more susceptible to colorimetric interference.

WST: Water soluble Tetrazolium salts are water soluble dyes for colorimetric detection developed to give different absorption spectra of the formed formazans. The advantage of WST over MTT is that they are reduced outside cells and on reacting with electron mediator 1-methoxy phenazine methosulfate (PMS), they yield a water soluble formazan. The additional solubilisation step as in case of MTT is not required as the water soluble formazan is the product which increases the sensitivity of the assay.

It is important to keep in mind that reduction of tetrazolium salts may vary depending upon the specific conditions and a particular cell type. The cells such as thymocytes reduce MTT at a slow rate compared to cells which are rapidly dividing. MTT assays should be done in the dark since the MTT reagent is sensitive to light. These assays are rapid, cost effective, and convenient and require no harvesting of cells. These assays can also be used to measure cytotoxicity or cytostatic activity (shift from proliferation to quiescence) of treatment.

4.7. Lactate Dehydrogenase Assay

In a laboratory focused on cell-based research measurement of cell cytotoxicity is an essential technique. This monitoring allows for the optimization of cell culture conditions. More importantly, the cytotoxic nature of anticancer compounds in toxicology testing, the toxicity of therapeutic chemicals in drug screening, and cell-mediated cytotoxicity can all be assessed through this assay-based approach. Cell cytotoxicity characteristics include loss of cellular metabolic activity or cell membrane integrity. The leakage of cellular components from dead cells into the culture media is one of the methods for assessing cell viability. Lactate dehydrogenase, an oxidoreductase that transfers a hydride from one molecule to another has long been favoured as a marker of cell death. Lactate dehydrogenase (LDH) is a cytosolic enzyme present in nearly all living cells (animals, plants, and prokaryotes). The enzyme is involved in the inter conversion of final product of glycolysis *i.e.* pyruvate to lactate under low supply of Oxygen. The reaction takes place with the concomitant reduction of NADH to NAD+. The LDH enzyme is inhibited by feedback mechanism that is the high concentrations of lactate inhibit the activity of enzyme [10].

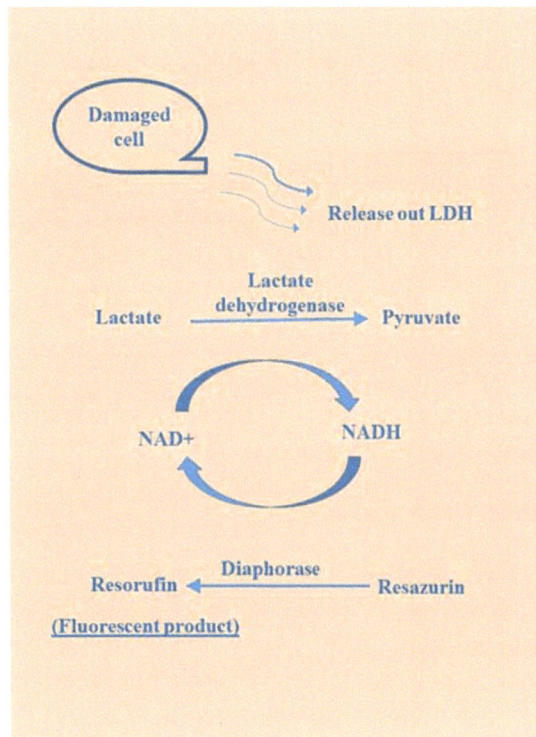

Fig. (4.3). Schematic representation of Lactate dehydrogenase assay.

LDH is released into culture medium upon cell death due to damage of plasma membrane. The increase of the LDH activity in culture supernatant is proportional to the number of lysed cells. Since LDH is a fairly stable enzyme, LDH has been widely used to evaluate the presence of damage and toxicity of tissue and cells. (Fig. **4.3**). LDH assay is an enzymatic cell death assay, performed by assessing cytosolic LDH soluble enzyme released into the media due to damage of plasma membrane as a marker of dead cells. LDH catalyses the interconversion of NAD^+ to NADH in the presence of L-lactate, while the formation of NADH can be quantified in a coupled reaction in which the tetrazolium salt is reduced to a red formazan product. The quantification of colored product is done spectrophotometrically at 490nm [11, 12].

Cell are seeded and besides test samples, positive and negative controls are included. Cultured cells are exposed to external treatments whose cytotoxic effect is to be evaluated. During experiments, culture medium is removed from plates at different time points like 6 h, 24 h, 48 h and transferred to 96 well plates. LDH substrate is added to each well containing culture media in ratio of 1:1 and incubated for 20 min at room temperature; the enzymatic reaction is arrested by adding stop solution to above mixture in ratio of 4:1. The LDH levels corresponding to maximum cell death are measured by treating one of the groups of cells with lysis buffer. Basal LDH levels are determined from control cells (without any treatment).

$$\% \text{ Relative Cytotoxicity} = \frac{\text{OD sample} - \text{OD negative control}}{\text{OD positive control} - \text{OD negative control}} \times 100$$

However, signal interference, signal quenching and test compound interference are among the disadvantages of the assay.

4.8. ATP Bioluminescent Assays

Cells growing in *in vitro* culture systems contain a relatively constant level of ATP to maintain homeostasis. However, the ability to synthesize ATP is lost during the process of cellular death. Thus, measuring the amount of ATP from the samples under most experimental conditions has been largely accepted as an authentic marker of the number of viable cells. The ATP assay is based on the property of luciferase enzyme to generate a luminescent signal [13]. The original protocol for the ATP assay involves an acid extraction step to degrade endogenous ATPase and to stabilise the amount of ATP present in the samples. This was followed by the neutralisation of the sample pH and the addition of luciferin and luciferase prepared from firefly. The signal from the reaction was a short lasting flash of light. However, the optimisation of the components of the luciferase assay resulted in the long lasting signal. The development of highly stable luciferase led to the main improvements in the ATP assay. The concentrations of enzyme, substrate and inhibitors were also optimised for providing flexibility in recording data from multiwell plates [14].

The ATP detection assay is one of the most sensitive microplate assays for detecting viable cells in culture. The low interference from test compounds in fluorescent assays and the speed of performing the assay provide advantage to ATP assay over other assays. Another advantage is the immediate lyses of cells on addition of reagents. The other assays require several hours of incubation with viable cells to convert the substrate to a colored product. The disadvantage is that the sample can't be used for other purposes as the assays system kills the cells. The other disadvantage is that the test compounds may inhibit luciferase activity giving false results, although reagent formulations have been designed to reduce the compound interference.

4.9. Clonogenic Assay

The transformation of cells is linked with certain phenotypic changes such as loss of contact inhibition and anchorage independence; cells grow over one another and form colonies. This property of transformed cells is known as clonogenicity. The normal cells when suspended in a medium like agar or agarose are unable to grow but transformed cell lines can grow and become anchorage-independent. The process, by which these changes occur, is believed to be closely associated to the process of *in vivo* carcinogenesis [15]. Puck and Marcus in 1956, for the first time described a cell culture technique for assessment of the clonogenicity of cells *in vitro* after treatment with external agents. This clonogenic assay has been used extensively for different studies with many types of cells, using improved complex culture media. The clonogenic assay determines the number of cells in a culture that retain the ability of a cell to proliferate indefinitely. The cells which retain the reproductive ability form a large colony (clone) and are said to be clonogenic. This assay was used to study the radiation effects on cells *in vitro* and is now used to assess the effect of chemotherapeutic agents, anti-angiogenic agents, cytokines and their receptors *etc*.

Clonogenic cell survival assay is carried out to understand the effect of external treatments on the ability of cells to proliferate indefinitely, thereby retaining its reproductive capacity to form a large colony. The colony formation is an anchorage independent growth assay in soft agar, considered the most stringent assay for assessing the degree of malignancy. In order to examine the effect of external treatment on contact inhibition and anchorage independence ability of cells *in vitro*, three-layer colony formation assay with a base layer consisting of 1% agar, a second layer containing cells with 0.4% agar and a third layer containing medium for growth is utilized. Different amounts of soft agar are required for culture dishes of different capacities as mentioned in Table (**4.1**).

Table 4.1. Suggested amounts for soft agar colony formation assay.

Culture Dish	96 well	48 well	24 well	6 well	35 mm	60 mm	100 mm
Base and Top Agar Volume (mL/well)	0.1	0.2	0.5	1.0	1.5	3.0	5.0
Cells/Well	500	1,000	1,250	2,500	5,000	7,500	12,500
Media volume (ml/Well) for feeding	0.05	0.1	0.25	0.5	0.75	1.5	2.5

Clonogenic assay is performed usually in 30mm or 60 mm cell culture dishes. In this assay, cells from actively growing adherent cultures are trypsinized so as to detach the cells from the substratum. The suspension of cells is prepared and the number of cells per millilitre are counted using hemocytometer. From the stock cultures, fixed numbers of cells are seeded into a dish with enriched culture medium in 0.4% agar. The agar-cell mixture is plated over a bottom layer of 1% agar in enriched complete medium plated in a culture dish. An overlay of complete culture media is added, cells are subjected to treatment and incubated for 24 h and subsequently media is discarded and fresh media is added. Treated cells along with appropriate controls are incubated in soft agar medium for 14-21 days with periodic replenishment of culture media. Following this incubation period, media is discarded and dishes are rinsed carefully with buffered saline. After washing 2-3 ml of a mixture of 6% glutaraldehyde and 0.5% crystal violet is added to each dish and incubated for at least 30 min. The glutaraldehyde crystal violet mixture is carefully removed and dishes are rinsed with tap water, and left to dry in normal air at room temperature. In order to quantify the number of colonies formed per well, colonies are counted in 10 random fields (Fig. **4.4**). Colony formation counts are performed with an automatic image-analysis system such as Image Jv1.42l. Cells which remain single and show evidence of nuclear deterioration and cells which go through one or two divisions and form small colonies of just a few cells are not counted and scored as dead [16]. Cells which form large colonies indicate the number of cells which have survived the treatment and have retained the ability to reproduce indefinitely. Plating efficiency [PE] of control cells, that is, the fraction of colonies among cells not affected by treatment is determined as:

The surviving fraction [SF] of cells after treatment is calculated taking into account the PE of control cells. Statistically the difference between control and treated groups is analyzed from data derived from three different experiments using statistical software.

$$PE = \frac{Number\ of\ colonies\ formed}{Number\ of\ cells\ seeded} \times 100\%$$

$$SF = \frac{Number\ of\ colonies\ formed\ after\ treatment}{Number\ of\ cells\ seeded} \times PE$$

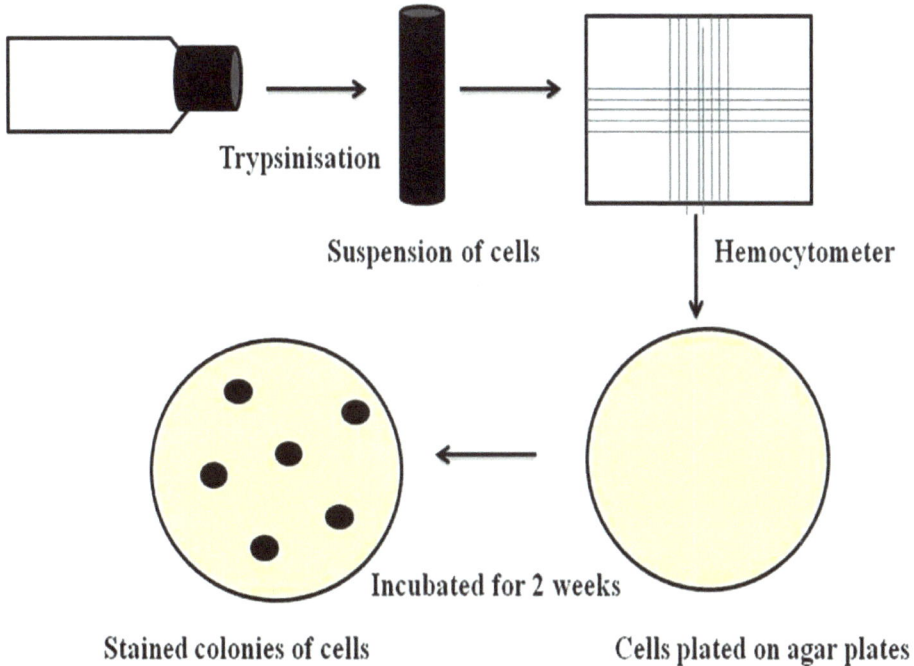

Fig. (4.4). Schematic representation of Clonogenic assay.

4.10. DNA Synthesis as a Marker for Cell Proliferation

One of the important methods for analysing cell proliferation is the measurement of DNA synthesis. In these assays, the incorporation of labelled nucleotide precursors is monitored and quantified. The amount of labelled precursor incorporated is directly proportional to the extent of cell division occurring in the culture. BrdU cell proliferation assay is one of the assay that utilises the halogenated nucleotide *i.e.* bromodeoxyuridine (BrdU) for labelling nascent DNA. When cells are cultured with labelled medium, the proliferating cells incorporate BrdU in place of thymine. The labelled medium is then removed and the cells are fixed and denaturing solution is used to denature DNA. The detection is done by using antibody against BrdU and finally the signal is quantified [17].

CONCLUDING REMARKS

The assays to check the cell viability, toxicity and proliferation are frequently carried out in *in vitro* culture models. They are involved in studying the toxicity of drugs, screening of drug targets for different diseases, anti-proliferative activity of natural and synthetic compounds, effect of gene-transfections, effect of gene silencing/over-expression/deletion and many more. In this chapter, we have discussed the principles, advantages and disadvantages of most of these assays and the differences amongst them. However, the selection of a particular assay depends upon the need of your research and specifically to the experimental settings.

CONSENT FOR PUBLICATION

Not applicable

CONFLICT OF INTEREST

None Declare

ACKNOWLEDGEMENTS

None Declare

REFERENCES

[1]   Strober W. Trypan blue exclusion test of cell viability. Curr Protoc Immunol 2001; 3(Appendix): 3B.
        [PMID: 18432654]

[2]   Ates G, Vanhaecke T, Rogiers V, Rodrigues RM. Assaaying cellular viability using the neutral red
        uptake assay. Methods Mol Biol 2017; 1601: 19-26.
        [http://dx.doi.org/10.1007/978-1-4939-6960-9_2] [PMID: 28470514]

[3]   Borenfreund E, Puerner JA. Toxicity determined *in vitro* by morphological alterations and neutral red

absorption. Toxicol Lett 1985; 24(2-3): 119-24.
[http://dx.doi.org/10.1016/0378-4274(85)90046-3] [PMID: 3983963]

[4]   Karászi E, Jakab K, Homolya L, *et al.* Calcein assay for multidrug resistance reliably predicts therapy response and survival rate in acute myeloid leukaemia. Br J Haematol 2001; 112(2): 308-14.
[http://dx.doi.org/10.1046/j.1365-2141.2001.02554.x] [PMID: 11167823]

[5]   Parish CR. Fluorescent dyes for lymphocyte migration and proliferation studies. Immunol Cell Biol 1999; 77(6): 499-508.
[http://dx.doi.org/10.1046/j.1440-1711.1999.00877.x] [PMID: 10571670]

[6]   Valla V, Christianopoulou MB, Kojic V, *et al.* Synthesis, spectroscopy and *in vitro* cytotoxycity of new hydroxy- anthraquinonatotriorganotin compounds. Synth React Inorg Met-Org Nano-Met Chem 2007; 37: 41-51.
[http://dx.doi.org/10.1080/15533170601172450]

[7]   Mosmann T. Rapid colorimetric assay for cellular growth and survival: application to proliferation and cytotoxicity assays. J Immunol Methods 1983; 65(1-2): 55-63.
[http://dx.doi.org/10.1016/0022-1759(83)90303-4] [PMID: 6606682]

[8]   Cory AH, Owen TC, Barltrop JA, Cory JG. Use of an aqueous soluble tetrazolium/formazan assay for cell growth assays in culture. Cancer Commun 1991; 3(7): 207-12.
[http://dx.doi.org/10.3727/095535491820873191] [PMID: 1867954]

[9]   Ishiyama M, Shiga M, Sasamoto K, *et al.* A new sulfonated tetrazolium salt that produces a highly water soluble formazan dye. Chem Pharm Bull (Tokyo) 1993; 41: 1118-22.
[http://dx.doi.org/10.1248/cpb.41.1118]

[10]   Decker T, Lohmann-Matthes ML. A quick and simple method for the quantitation of lactate dehydrogenase release in measurements of cellular cytotoxicity and tumor necrosis factor (TNF) activity. J Immunol Methods 1988; 115(1): 61-9.
[http://dx.doi.org/10.1016/0022-1759(88)90310-9] [PMID: 3192948]

[11]   Anderson BM, Noble C Jr. *In vitro* inhibition of lactate dehydrogenases by kepone. J Agric Food Chem 1976; 25(1): 28-31.
[http://dx.doi.org/10.1021/jf60209a004] [PMID: 63476]

[12]   Sanders GT, van der Neut E, van Straalen JP. Inhibition of lactate dehydrogenase isoenzymes by sodium perchlorate evaluated. Clin Chem 1990; 36(11): 1964-6.
[PMID: 2173649]

[13]   Crouch SP, Kozlowski R, Slater KJ, Fletcher J. The use of ATP bioluminescence as a measure of cell proliferation and cytotoxicity. J Immunol Methods 1993; 160(1): 81-8.
[http://dx.doi.org/10.1016/0022-1759(93)90011-U] [PMID: 7680699]

[14]   Hall MP, Gruber MG, Hannah RR, *et al.* Stabilization of firefly luciferase using directed evolution. Bioluminescence and chemiluminescence –perspectives for the 21st century. New York: CRC Press 1998.

[15]   Munshi A, Hobbs M, Meyn RE. Clonogenic cell survival assay. Methods Mol Med 2005; 110: 21-8.
[PMID: 15901923]

[16]   Franken NA, Rodermond HM, Stap J, Haveman J, van Bree C. Clonogenic assay of cells *in vitro.* Nat Protoc 2006; 1(5): 2315-9.
[http://dx.doi.org/10.1038/nprot.2006.339] [PMID: 17406473]

[17]   Bosq J, Bourhis J. [Bromodeoxyuridine (BrdU). Analysis of cellular proliferation]. Ann Pathol 1997; 17(3): 171-8.
[PMID: 9296576]

CHAPTER 5

Reactive Oxygen Species (ROS)

Abstract: The reactive oxygen species produced endogenously are essential to life, being involved in several biological functions. However, when produced at higher levels, these reactive species become highly harmful, causing oxidative stress through the oxidation of biomolecules, leading to cellular damage. The study in the field of ROS associated biological functions and/or deleterious effects requires new sensitive and specific tools in order to enable a deeper insight on its action mechanisms. Here, we discuss several methods related to the detection of reactive oxygen species.

Keywords: Catalase, Fluorescence, Fluorescent assays, Free radicals, Hydrogen peroxide, Mitochondria, Superoxide dismutase.

INTRODUCTION

The reactive molecules and the free radicals derived from molecular oxygen are referred to as reactive oxygen species. The unpaired electrons in the atomic oxygen make it susceptible to radical formation and the sequential reduction of oxygen leads to the formation of a number of reactive oxygen species like superoxide, hydrogen peroxide, hydroxyl radical, hydroxyl ion, nitric oxide, *etc* [1]. The endogenous sources of ROS include mitochondria, peroxisomes, cytochrome P450 metabolism and inflammatory cell activation. The ROS is primarily produced by the phagocytic cells where the process is catalysed by the action of NADPH oxidase. Though several enzymes are involved in the production of ROS moieties; however NADPH oxidase is the most significant [2]. The reactive oxygen species have role in apoptosis, gene expression and triggering of cell signalling cascades; however, their increased level in the cellular system has potential to result in deleterious events [3]. The detoxification of ROS is necessary for the survival of aerobic life forms and thus a number of cellular defence mechanisms have been developed to maintain a balance between the production and removal of ROS. Superoxide Dismutase is an enzyme that catalyses the conversion of superoxide anions into hydrogen peroxide and oxygen. The hydrogen peroxide thus produced is then catalysed by other enzyme (catalase) into water and oxygen. The Glutathione peroxidase is another group of enzymes which also catalyses the degradation of hydrogen peroxide and organic peroxides. Glutathione is among the non-enzymatic small molecule that plays a key role in the detoxification mechanisms. Glutathione is a tripeptide (glutamyl-cysteinyl-glycine), possessing an exposed sulphydryl group. Glutathione gets oxidised on reacting with ROS and the reduced form is regenerated by NADPH dependent reductase. A dynamic indicator of oxidative stress is the ratio of oxidised form of glutathione (GSSG) to the reduced form (GSH). Ascorbic acid, a water soluble molecule is also capable of reducing ROS, while alpha-tocopherol, a lipid soluble molecule has antioxidant role in membranes [4].

The strength, duration and context of ROS determine the effect on cellular processes. These reactive species and the antioxidant enzymes in conjugation play a key role in turning on or off redox signalling cascades. Nitric oxide acts as cellular messenger and co-ordinates several effects. The apoptotic mechanisms are triggered due to increased level of ROS. Stress activates p21 proteins and blocks cell cycle progression [5]. Exposure to hydrogen peroxide or nitric oxide results in the dephosphorylation of Retinoblastoma (Rb) and thus cell cycle arrests. ROS is believed to contribute to the development of age related diseases, cancer, atherosclerosis, diabetes and neurodegenerative diseases. Keeping in view the role of ROS in various cellular processes, the accurate measurement of these reactive species is necessary. However, the measurement of ROS is dependent upon the analytic target and the reactive species in question. In the present section we will discuss some of the assays for the measurement of ROS in *in vivo* conditions.

5.1. Detection of Lipid Peroxidation

The measurement of lipid peroxidation is one of the widely used indicators of free radical formation. Since the prime targets of free radicals are the unsaturated fatty acids present on the cellular membranes. The free radicals attack the hydrogen moiety of unsaturated fatty acid to form H_2O leaving an unpaired electron on the unsaturated carbon, which then captures oxygen to form a peroxy radical (Fig. **5.1**). The lipid peroxides formed are unstable and they decompose to form series of reactive carbonyl compounds such as malondialdehyde. Measurement of reactive oxygen species by this method has been widely used since long times and it is based on the reaction of MDA with Thiobarbituric acid [6]. The reaction takes place under acidic conditions and at a temperature range of 90-100C, and results in the formation of adduct that can be measured.

5.2. Glutathione Assay

Glutathione is a tripeptide, consisting of cysteine, glycine, and glutamate. A gamma peptide linkage exists between the carboxyl group of the glutamate side chain and the amine group of cysteine, while the carboxyl group of cysteine is attached by normal peptide linkage to a glycine. It has a sulfhydryl (SH) group on the cysteinyl portion, which accounts for its strong electron-donating character.

On loss of electrons molecules become oxidized, and as such two molecules are dimerized reversibly by a disulphide bridge to form glutathione disulphide or oxidized glutathione (GSSG) [7]. Reduced glutathione is the major free thiol and key antioxidant present in the living cells. It's involved in several biological processes like removal of hydro peroxides, detoxification of xenobiotics and maintenance of oxidation state. About 90-95% of total glutathione is present in the reduced form and rest in the oxidised form. The reduced glutathione is regenerated by NADPH dependent reductase into its oxidised form. The ratio of the reduced to the oxidised form is a critical indicator of the oxidative stress. Several assays have been designed for measuring glutathione in test samples.

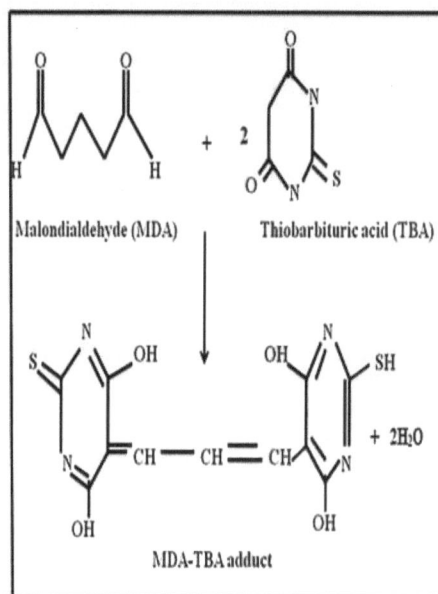

Fig. (5.1). Molecular structure and formation of MDA-TBA adduct.

5-5'-dithiobis [2-nitrobenzoic acid] (DTNB) or Ellman's Reagent is used to quantitate thiols, it is commonly used to measure glutathione [8]. It reacts with GSH to form 5-thionitrobenzoic acid (TNB) and GS-TNB, a chromophore to give a yellow colored adduct 5-thio-2 nitrobenzoic acid which can be quantified at 412nm (Fig. **5.2**). DTNB reacts with GSH to produce a conjugate and TNB anion that can be detected by fluorescence or absorbance. To measure total GSH, a recycling assay is used in which GSH reacts with the conjugate producing GSSG and another molecule of TNB, which increases fluorescence or absorbance. Glutathione reductase further reduces GS-TNB in presence of β-nicotinamide adenine dinucleotide phosphate (NADPH), releasing a second TMB molecule and recycling the GSH. Any oxidized GSH (GSSG) initially present in the reaction mixture or formed from the mixed disulphide reaction of GSH with GS-TNB is rapidly reduced to GSH.

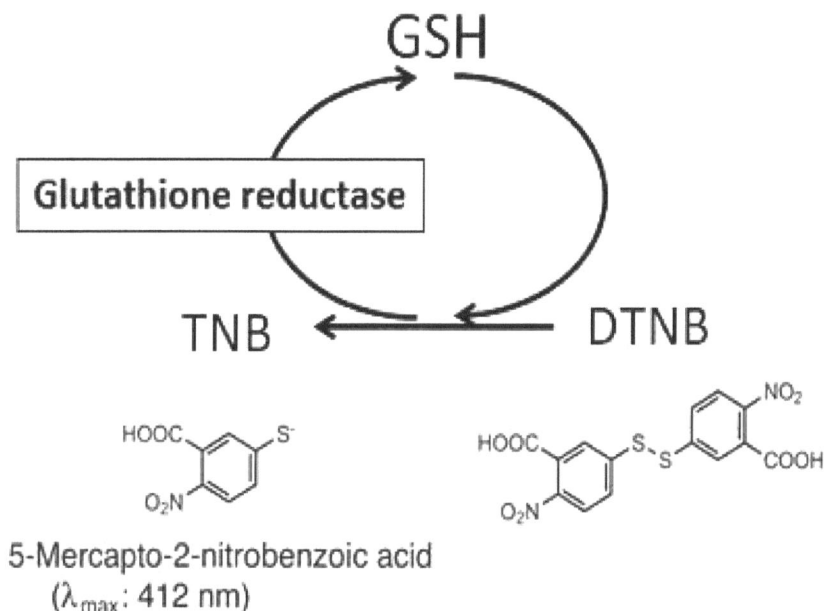

Fig. (5.2). Reaction catalysed by Glutathione Reductase.

Bimane such as Monochlorobimane is another approach used to quantify glutathione [9]. This method is more sensitive than other methods of detection. Cells are lysed and thiols are reduced with dithiothreitol and labelled by bimane. Bimane becomes fluorescent after binding to GSH. Flouresence quantification is done by confocal laser scanning which relies on measuring the rates of fluorescence changes. The thiols can be separated by HPLC and the fluorescence quantified with a fluorescence detector.

5-Chloromethylfluorescein diacetate (CMFDA) is a cell-permeable green fluorescent compound used as a glutathione probe [10]. Inside cell it transforms intracellularly into a cell-impermeant, brightly fluorescent product. It is non-toxic and stable at physiological pH. Fluorescence of CMFDA increases due to the hydrolysis of the acetate groups inside cells as such fluorescence increase does not reflect the reaction Although CMFDA may react with glutathione in cells, unlike bimane, whose fluorescence increases upon reacting with glutathione.

Bimane based probes are based on irreversible chemical reactions with glutathione, which renders these probes incapable of monitoring the glutathione in real-time. So a probe based on reversible reaction has been reported recently,

Thiol Quant Green (TQG). Thiol Quant Green is highly sensitive and performs high resolution measurements of glutathione levels in single cells using a confocal microscope. It can also be applied in flow cytometry to perform bulk measurements.

Another approach, which allows measurement of the glutathione redox potential at a high spatial and temporal resolution in living cells is based on redox imaging using the redox-sensitive green fluorescent protein (roGFP) or redox sensitive yellow fluorescent protein (rxYFP) GSSG because it's very low physiological concentration is difficult to measure accurately unless the procedure is carefully executed and monitored and the occurrence of interfering compounds is properly addressed. Similarly, the amount of reduced glutathione can be determined by using a luciferin derivative in conjugation with glutathione-S-transferase which results in the generation of a luminescent signal when luciferin is added.

5.3. Superoxide Assay

The superoxide anion, systematically known as dioxide or hyperoxide, with the chemical formula O_2^- is a short lived free radical generated from oxygen by the addition of an electron. In cell superoxide is normally produced as a defence tool in phagocytes by the enzyme NADPH oxidase for use in oxygen-dependent killing mechanisms of pathogens. However, it is also produced as a by product of various cellular reactions such as mitochondrial respiration, cellular enzymes like NADPH oxidase, xanthine oxidase, environmental factors such as cigarette smoke, UV light, gamma radiation, pollutants. The superoxide anions cause damage to lipids, proteins, DNA and thus can initiate numerous diseases including diabetes, rheumatoid arthritis, cancer, liver damage *etc*.

Superoxide is highly reactive and has a very short half-life as a result its quantification is a very difficult task. Based on this fact assays are designed such that superoxide is converted into a stable product which is later assayed to give an indirect measurement of superoxide. On this approach an assay is based in which superoxide is converted to hydrogen peroxide, which is relatively stable. Hydrogen peroxide is then assayed by a fluorimetric method. Another property of superoxide is that it has a strong Electron paramagnetic resonance (EPR), which can be used to detect superoxide directly. Spin traps, a series of tool compounds have been developed that can react with superoxide, forming a meta-stable radical which can be more readily detected by EPR. Superoxide spin trapping was initially carried out with 5,5-Dimethyl-1-Pyrroline-N-Oxide (DMPO), but, more recently, phosphorus derivatives with improved half-lives, such as DEPPMPO and DIPPMPO, have become more widely used. However, practically this assay can be done only under non-physiological conditions, such as high pH.

Alternatively the detection of superoxide ions can be done based on the interaction with some other compounds to give a measurable result. Colorimetrically, the rate of superoxide formation is determined by the reduction of ferricytochrome c to ferrocytochrome c and the reaction is monitored at 550nm. However the reaction is not completely specific for superoxide. Other method involves the use of aconitase, that catalyses the conversion of citrate to isocitrate. The enzyme aconitase is inactivated by the addition of superoxide anions and thus the degree of enzyme inactivation determines the superoxide concentrations. In this assay, the enzyme activity is measured by using fixed amount of substrate and the activity is monitored at 240nm. The sensitivity of assay can further be increased by coupling with other assay where the aconitase product *i.e.* isocitrate is converted to alpha-ketoglutarate by NADP+ dependent isocitrate dehydrogenase. In this case the NADPH is monitored colorimetrically at 540nm.

Apart from colorimetric method some chemilumniscent reactions have also been used. Lucigenin and coelenterazine are used as chemilumniscent substrate [11]. The fluorogenic sensors for superoxide radicals are the hydrocyanine dyes. Upon oxidation, fluorescence intensity of these dyes increases about 100 fold.

5.4. Nitric Oxide Detection Assay

Nitric oxide is formed from the amino acid L-arginine and the reaction is catalysed by nitric oxide synthase (Fig. **5.3**). NO acts a unique intracellular and extracellular messenger and is involved in physiological process like neurotransmission, vasodilation, cytotoxicity and inflammation. Griess assay is one of the commonly used methods for the estimation of NO and it involves the indirect measurement of its composition products, nitrate and nitrite [12]. In this method, first nitrate is converted into nitrite by nitrate reductase. Subsequently, in a two-step process, the total of nitrate and nitrite is determined. Nitrite in presence of hydrogen ions forms nitrous acid which on reacting with sulphanilamide produces a diazonium ion. The ion formed couples to N-(1-napthyl) ethylene-diamine to form chromophore which absorbs at 543nm. For the estimation of nitrite only, a parallel assay is carried out where samples are not reduced. The nitrate levels are then calculated by subtracting nitrite levels from the total. This assay is easy to perform and inexpensive.

Fluorimetric assays have been developed for increasing the sensitivity of Griess assay. The reaction is dependent on dinitrogen trioxide formed by the acidification of nitrite. One of the commonly used method employs 2,3-diaminonapthalene (non-fluorescent) to react with nitrous acid to form 2,3 napthotriazole which is a fluorescent molecule having an excitation wavelength of 375nm.

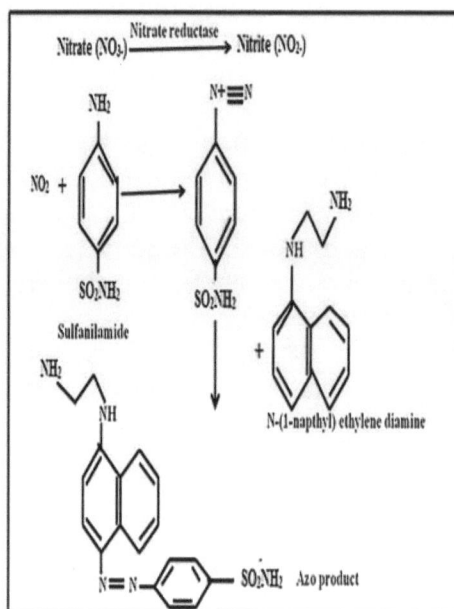

Fig. (5.3). Determination of nitrate by Griess assay.

5.5. Hydrogen Peroxide Detection Assay

Hydrogen peroxide is the simplest peroxide with an oxygen–oxygen single bond. Hydrogen peroxide, dominated by the nature of its unstable peroxide bond slowly decomposes in the presence of base or a catalyst. Enzymes that use or decompose hydrogen peroxide are classified as peroxidases. Hydrogen peroxide is found in biological systems including the human body. It is a metabolic by-product of aerobic respiration in mitochondria and is produced as a defence mechanism within phagocytes. It has long have been associated with various biological processes and disorders including aging and cancer. Hydrogen peroxide as a result can serve as an indicator of oxidative stress. Measurement of this reactive species may help to determine how it serves as a key regulator for a number of oxidative stress-related states and can be a prognostic and diagnostic marker in various diseases. Hydrogen peroxide has important roles as a signalling molecule in the regulation of a wide variety of biological processes functioning through NF-κB and other factors. Hydrogen peroxide readily decomposes into a hydroxyl radical and superoxide radical which in turn readily reacts with and damages vital cellular components, especially mitochondria. The higher concentrations of H_2O_2 inhibit oxidation of DNA, lipids and proteins leading to mutagenesis and cell death [13, 14]. The role of H_2O_2 has been implicated in a number of pathological conditions such as asthma, diabetes, neurodegenerative diseases and thus appropriate methods should be used to quantify the H_2O_2 concentration. A number of fluorogenic substrates in conjugation with horseradish peroxide (HRP) have been used for the detection of hydrogen peroxides. Diacetyldichloro-fluorescein, homovanillic acid and amplex red are the commonly used substrates (Fig. **5.4**). In these assays, the increasing concentrations of H_2O_2 results in the increased production of fluorescent products *e.g.* the hydrogen peroxide oxidises amplex red in presence of horseradish peroxidase into a colored compound (resorufin) that can be detected colorimetrically at 570nm or fluorimetrically using excitation of 570nm and emission 585nm [15]. Some colorimetric substrates such as tetramethylbenzidine and phenol red have also been used in conjugation with HRP to measure hydrogen peroxide concentrations. Though the fluorescent assays are more sensitive than the colorimetric assays but care should be taken while carrying out the fluorescent assays. The thiols present in the cellular systems can serve as substrates for HRP and the endogenous catalase can reduce the actual amount of H_2O_2.

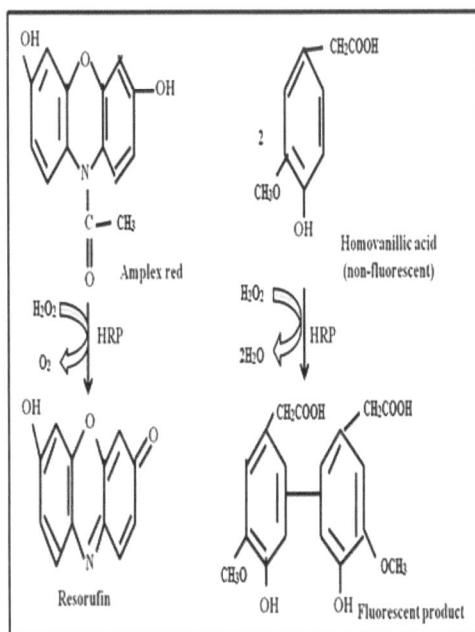

Fig. (5.4). Detection of hydrogen peroxides by the conversion of substrates (amplex red and homovanillic acid) into detectable products.

One of the extensively used assays for the quantitation of hydrogen peroxide involves the oxidation of 2'-7' dichlorofluorescein (H2DCF) to. 2'-7' dichloro-fluorescein (DCF) [16]. First the diacetate form of H2DCF and its acetomethyl ester H2DCFDA enter into the cells. Inside the cells, they are cleaved by cellular esterases resulting in a charged molecule. Then the charged molecule (H2DCF) gets oxidised by ROS into a highly fluorescent compound 2'7'dichlorodihydrofluorescein (DCF) which is measured at 498nm for excitation and 522nm for emission.

CONCLUDING REMARKS

We have discussed several assays for the measurement of reactive oxygen species in *in vitro* conditions. The choice of method used to measure a particular process depends on *how* the method works, *how* quantitative it can be and *what* is likely to confound it. By taking these points into consideration, erroneous interpretation can be minimised which will be helpful in better understanding and diagnosis of numerous diseases.

CONSENT FOR PUBLICATION

Not applicable

CONFLICT OF INTEREST

None Declare

ACKNOWLEDGEMENTS

None Declare

REFERENCES

[1]   Buechter DD. Free radicals and oxygen toxicity. Pharm Res 1988; 5(5): 253-60.
        [http://dx.doi.org/10.1023/A:1015914418014] [PMID: 3072554]

[2]   Waris G, Ahsan H. Reactive oxygen species: role in the development of cancer and various chronic conditions. J Carcinog 2006; 5: 14.
        [http://dx.doi.org/10.1186/1477-3163-5-14] [PMID: 16689993]

[3]   Hancock JT, Desikan R, Neill SJ. Role of reactive oxygen species in cell signalling pathways. Biochem Soc Trans 2001; 29(Pt 2): 345-50.
        [http://dx.doi.org/10.1042/bst0290345] [PMID: 11356180]

[4]   Matés JM. Effects of antioxidant enzymes in the molecular control of reactive oxygen species toxicology. Toxicology 2000; 153(1-3): 83-104.
        [http://dx.doi.org/10.1016/S0300-483X(00)00306-1] [PMID: 11090949]

[5]   Gartel AL, Radhakrishnan SK. Lost in transcription: p21 repression, mechanisms, and consequences. Cancer Res 2005; 65(10): 3980-5.
        [http://dx.doi.org/10.1158/0008-5472.CAN-04-3995] [PMID: 15899785]

[6]   Janero DR. Malondialdehyde and thiobarbituric acid-reactivity as diagnostic indices of lipid peroxidation and peroxidative tissue injury. Free Radic Biol Med 1990; 9(6): 515-40.
        [http://dx.doi.org/10.1016/0891-5849(90)90131-2] [PMID: 2079232]

[7]   Jones DP. Redox potential of GSH/GSSG couple: assay and biological significance. Methods Enzymol 2002; 348: 93-112.
        [http://dx.doi.org/10.1016/S0076-6879(02)48630-2] [PMID: 11885298]

[8]   Ellman GL. Tissue sulfhydryl groups. Arch Biochem Biophys 1959; 82(1): 70-7.

[http://dx.doi.org/10.1016/0003-9861(59)90090-6] [PMID: 13650640]

[9]   Kamencic H, Lyon A, Paterson PG, Juurlink BH. Monochlorobimane fluorometric method to measure tissue glutathione. Anal Biochem 2000; 286(1): 35-7.
[http://dx.doi.org/10.1006/abio.2000.4765] [PMID: 11038270]

[10]   Sebastià J, Cristòfol R, Martín M, Rodríguez-Farré E, Sanfeliu C. Evaluation of fluorescent dyes for measuring intracellular glutathione content in primary cultures of human neurons and neuroblastoma SH-SY5Y. Cytometry A 2003; 51(1): 16-25.
[http://dx.doi.org/10.1002/cyto.a.10003] [PMID: 12500301]

[11]   Barbacanne MA, Souchard JP, Darblade B, *et al.* Detection of superoxide anion released extracellularly by endothelial cells using cytochrome c reduction, ESR, fluorescence and lucigenin-enhanced chemiluminescence techniques. Free Radic Biol Med 2000; 29(5): 388-96.
[http://dx.doi.org/10.1016/S0891-5849(00)00336-1] [PMID: 11020659]

[12]   Green LC, Wagner DA, Glogowski J, Skipper PL, Wishnok JS, Tannenbaum SR. Analysis of nitrate, nitrite, and [15N]nitrate in biological fluids. Anal Biochem 1982; 126(1): 131-8.
[http://dx.doi.org/10.1016/0003-2697(82)90118-X] [PMID: 7181105]

[13]   Ruch W, Cooper PH, Baggiolini M. Assay of H_2O_2 production by macrophages and neutrophils with homovanillic acid and horse-radish peroxidase. J Immunol Methods 1983; 63(3): 347-57.
[http://dx.doi.org/10.1016/S0022-1759(83)80008-8] [PMID: 6631014]

[14]   Zhou M, Diwu Z, Panchuk-Voloshina N, Haugland RP. A stable nonfluorescent derivative of resorufin for the fluorometric determination of trace hydrogen peroxide: applications in detecting the activity of phagocyte NADPH oxidase and other oxidases. Anal Biochem 1997; 253(2): 162-8.
[http://dx.doi.org/10.1006/abio.1997.2391] [PMID: 9367498]

[15]   Reszka KJ, Wagner BA, Burns CP, Britigan BE. Effects of peroxidase substrates on the Amplex red/peroxidase assay: antioxidant properties of anthracyclines. Anal Biochem 2005; 342(2): 327-37.
[http://dx.doi.org/10.1016/j.ab.2005.04.017] [PMID: 15913534]

[16]   Rothe G, Valet G. Flow cytometric analysis of respiratory burst activity in phagocytes with hydroethidine and 2',7'-dichlorofluorescin. J Leukoc Biol 1990; 47(5): 440-8.
[http://dx.doi.org/10.1002/jlb.47.5.440] [PMID: 2159514]

CHAPTER 6

Protein-Protein Interactions

Abstract: Protein-protein interactions are central to every cellular process like DNA replication, transcription, translation, splicing, signal transduction and cell cycle control. To elucidate the function of a gene, we need to determine the function of the gene's encoded protein product. Identifying the role of a protein needs understanding about the proteins that interact with each other for functioning of a particular biological pathway. Assessment about function of a protein can be made by understanding protein-protein interaction studies. These inferences are based on the fact that the role of unknown proteins may be elucidated if captured through their interaction with a protein target of known function. Thus in this chapter, we attempt to summarize different *in vitro* methods used to identify proteins- protein interactions and to assess the strengths of these interactions.

Keywords: Agarose Beads, Co-Immunoprecipitation, Far Western Blotting, Monoclonal Antibody, Protein A, Pull Down Assay, SDS-PAGE, Sepharose.

INTRODUCTION

The Protein protein interactions (PPIs) which arise as a result of non-covalent contacts between the residues of side chains play a crucial role in the cellular processes including cell to cell interactions and regulatory control over metabolism and development [1]. The 30,000 genes in the human beings through a sequence of gene splicing events and post translational modifications are speculated to produce 1 into 10^6 proteins. The proteins can act either in isolation or in complexes with other proteins; however, the function of proteins in complexes or networks is necessary to co-ordinate the biological networks that impact cellular structure and function. The PPIs are essential in protein folding, translation, signalling, transcription, cell-cycle control and differentiation and post-translational modifications. The PPIs are categorised as stable or transient. The Stable type of interactions is those linked with proteins that are captured as multi-subunit complexes like in case of Haemoglobin and Core RNA polymerase. The transient interactions are short term in nature and require specific set of conditions that lead the interaction. They can be further classified as strong or weak and fast or slow. The transient interactions among proteins play a pivotal role in the recruitment and assembly of the transcription factors to specific promo-

Taseen Gul, Henah Mehraj Balkhi & Ehtishamul Haq

ters, the folding of native proteins mediated by chaperonins, the transport of proteins across membranes, individual steps of the translation cycle, and the assembly and breakdown of sub-cellular structures during the cell cycle. The consequences of protein-protein interactions include altering the kinetic features of enzymes. This may occur as a result of minor changes at the level of substrate binding or modifications in allosteric regulation. The interaction may serve an important role by regulating an upstream or a downstream action. This can lead to changes in the specificity of a protein for its substrate or they can create new binding site for small effector molecules and can even inactivate a protein [2].

The widely adopted *in vitro* methods for the discovery of protein interaction are based on the natural affinity of interacting binding partners [3, 4]. Some of them can be performed with basic laboratory skills whereas others require high expertise and a considerable investment in specialized instrumentation. The *in vitro* methods for protein interaction analysis include a broad range of techniques and some of them are discussed below:

6.1. Co-Immunoprecipitation

The concept of co-immunoprecipitation (co-IP) is based on the immunoprecipitation (IP) and to gain an understanding of the principles involved, the technique of immunoprecipitation will be discussed first [5]. IP is one of the most commonly used methods for protein detection and purification. The principle of an IP is very simple and involves isolating a specific protein from the mixture of proteins with the help of an antibody that has specific binding site for that particular protein. Antibody either monoclonal or polyclonal is used against a specific target antigen and allowed to form an immune complex with that target epitope in a sample. There are two alternate ways to start the immunoprecipitation procedure. In one approach, an antibody is first pre-immobilized on agarose or magnetic beads, and then treated with a cell lysate containing the target protein. In the second approach, first antibody is incubated with cell lysates and allowed to form immune complexes and then the protein complexes are captured by the insoluble support. The process of retrieving immune complexes from the solution is termed as precipitation. Finally, bound immune complex are eluted from the support and components (antigen and antibody) are analyzed by SDS-PAGE and identified by western blotting as shown in Fig. (**6.1**).

Co-immunoprecipitation (Co-IP) is a classical technique carried out in the similar way as an IP, except that the target protein (bait) precipitated by the antibody, is used to co-precipitate a binding partner or protein complex (prey) from a lysates. The prey protein is bound to bait protein which in turn is bound to antibody that is immobilized on to the support [6]. The proteins and their binding partners isolated by immunoprecipitation are then subjected to SDS-PAGE and Western blot analysis. The protocol for IP/Co-IP is schematically shown in Fig. (**6.2**) and it includes the following steps:

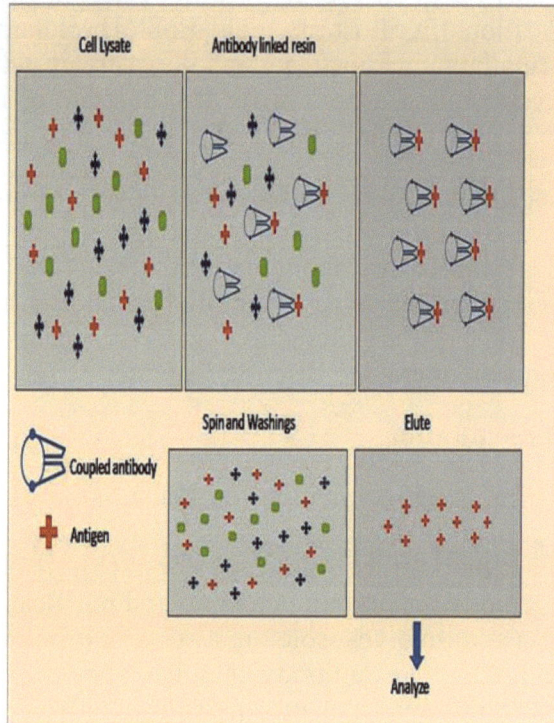

Fig. (6.1). Diagrammatic representation of a traditional Immuno-precipitation technique.

- Firstly, the cell lysates are prepared; the cells are resuspended in PBS and centrifuged at 3000 rpm for 5 minutes to collect the pellet. The cells are lysed under mild denaturation conditions by using Nonidet P-40 buffer (1%NP-40, Tris-Cl pH8.0, 1mM Poly Methyl Sulfonyl Fluoride, 150mM Sodium Chloride). The proteins are protected by proteases by using a cocktail of protease inhibitor which is added freshly to cell lysates. The Protease inhibitor cocktail is freshly added to the cell lysate. If the proteins are very sensitive to even mild detergents, then the cells are sonicated in a buffer with protease inhibitors. The cell lysate is cleared by centrifugation at 14,000 rpm for 10 minutes. The pellet is discarded and supernatant is transferred to vials and stored.
- The protein lysates are solubilised and precleared by incubating with protein G/A agarose suspension at room temperature for one hour. The suspension is centrifuged and the supernatant is used for the IP/Co-IP

- Take few microlitres (100-200μl) of the pre-cleared cell lysate and add 1-2 μg of the primary antibody and incubate overnight at 4°C.
- Then mix either protein A or G agarose beads (30 μl of 50% bead slurry) and Incubate for 1–3 hours at 4°C.
- Centrifuge at 3000 rpm for 30 seconds at 4°C. The pellet obtained is washed many times with 500 μl of cell lysis buffer (20 mM Tris-Cl pH 8.0, 1 mM Ethylene Diamine Tetra Acetic acid, and 200 mM Sodium Chloride, 1 mM PMSF). Maintain cold condition during the procedure.
- This is followed by resuspending the pellet with 20μl 1X SDS sample buffer and vortex before heating the IP mixture at 95–100°C for few minutes. Centrifuge the contents for 1 minute at 14,000 rpm to remove any in soluble matter.
- Analyze samples by SDS-PAGE gel (12–15%) followed by western blotting.

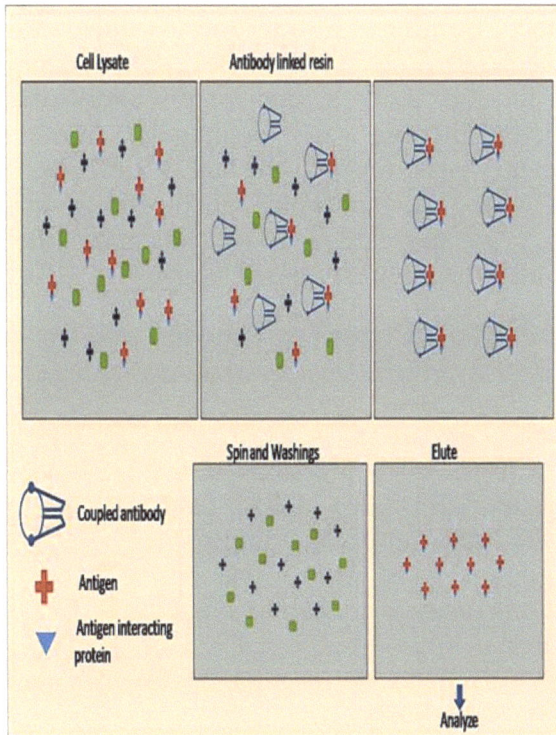

Fig. (6.2). Schematic Representation of a typical co-immunoprecipitation technique.

6.2. Pull Down Assays

It is an *in vitro* method for elucidating physical interactions among proteins and initial screening technique for identifying unknown protein-protein interactions. The assay is based on the concept of both affinity purification and immunoprecipitation. The assay involves the use of a tagged protein that is used to capture and "pull-down" a specific protein-binding partner. The tagged protein is immobilised by using ligand specific for the tag whereas the tagged protein acts as a secondary affinity support for capturing and purifying other proteins that interact with the bait protein [7]. The bait protein is treated with a variety of other protein sources that contain putative prey proteins. The source of prey protein depends on the aim of our experiment that is whether we are confirming already suspected protein-protein interactions or elucidating new protein-protein interactions.

Pull down assays for the confirmation of suspected interactions uses a prey protein that is expressed and purified from artificial protein expression system. This leads to the availability of large quantity of protein and avoids interference from interaction of the bait with other proteins present in the endogenous system. The purified proteins from *in vitro* transcription/translation reactions or protein expression system lysates are suitable prey protein sources for confirmatory studies. Whereas the pull down assays for discovery of unknown interactions involves native environment in which an unknown protein can interact with bait protein. The cellular lysates in which the bait is normally expressed, or complex biological fluids, where the bait would be functional are proper prey protein sources for discovery studies. Pull-down assays can provide higher resolution and selectivity than some other antibody-based assays [8].

For GST pull down assay, following steps are carried out to get the specific interaction partner of a protein of interest fused with GST.

- Clone the gene of the protein of interest in the GST expression vector in frame with the GST. Check the expression of the GST-fusion protein and once confirmed the correct expression of GST-fusion protein, more fusion protein is expressed from the scaled up BL21 cultures.
- The bacterial cells are lysed with NP-40 buffer followed by sonication. The GST-fusion protein containing bacterial extract is further processed and purified as discussed earlier.
- The cell extracts (in which we are looking for the interaction partner of the protein of interest) is prepared.
- The soluble protein fraction is pre-cleared by incubating with 50 ml of 50% (v/v) glutathione agarose suspension for 1h with continuous shaking.
- Equal amounts of pre-cleared protein extract is then incubated overnight at 4^0C with the purified GST-fusion protein and with the GST-only glutathione agarose beads (used as control) with continuous shaking.
- Next day beads are collected by centrifugation and washed 3-4 times with the GST wash buffer {20mMTris-Cl (pH8.0), 200mM NaCl, and 1mM EDTA,

0.5% NonidetP-40}.

- Finally, the beads (both GST-only beads and GST-protein of interest fusion beads) are resuspended in 1X SDS loading buffer and the eluted proteins are resolved on 10% SDS-PAGE gel and further processed. Depending upon the efficiency of interaction and the amount of the interacting protein present in the cells, the SDS PAGE is stained with commassie or silver stain to detect any protein band pulled down by the GST-fusion protein. The protein band is cut from the gel and indentified using mass spectrometry. Moreover, the identification of the suspected protein partner can be performed by immuno- bloting, using polyclonal antibody against the suspected interaction partner of the protein of interest.

6.3. Far-Western Blotting

This technique was initially developed to detect protein expression libraries but now it is used to decipher protein-protein interactions. Far Western Blotting has been used to know receptor-ligand interactions, screening libraries for interaction partners, studying effect of post translational on protein-protein interactions and to determine interaction sequences using artificially synthesised peptides as probes. The traditional far-Western analysis is similar to western blot analysis. Here a labelled or antibody-detectable "bait" protein is used to capture and detect the target "prey" protein on the membrane. The lysates containing the unknown prey protein is subjected to SDS or native PAGE and then transferred to a membrane. After binding to the membrane, the prey protein becomes accessible to probing. After transfer, the membrane is blocked and then probed with a known bait protein, which usually is applied in pure form. The incubation of bait and prey protein allows their interaction. The detection is done specific for the bait protein by using suitable detection system and the blots are analysed [9]. However, this technique was specialised to yield specific information about protein-protein interactions by modifying the design of bait proteins. The modified approach was used to elucidate the contact sites among subunits of a multi-subunit complex like interaction domains among the different subunits of of *E. coli* RNA polymerase were identified. The protein was expressed as a polyhistidine-tagged fusion, then partially cleaved and purified using a Ni^{2+} chelate affinity column. The polyhistidine-tagged fragments were separated by SDS-PAGE and transferred to a nitrocellulose membrane. The fragment-localized interaction domain was identified using a ^{32}P-labeled protein probe.

The far-western blotting technique must be performed with care and attention as there is a need to persevere the native conformation and interaction conditions for the proteins under study. The proteins in non-native and denatured conditions may not be able to interact leading to false results. Thus the preparative steps for this technique should be done carefully and use of experimental controls is necessary for appropriate validation.

CONCLUDING REMARKS

Protein interaction analysis uncovers unique, multiple and unforeseen functional roles of proteins. The previously unknown proteins may be discovered by their association with one or more proteins that are known. All the *in vitro* methods discussed in this spectrum have their place and can provide valuable insight to validate, characterize and confirm protein interactions. The affinity based methods are highly sensitive and are capable of detecting even weak interactions (10^{-5}M). The *in vitro* affinity based methods can be either direct (pull down or far-western analysis) or indirect as incise of co-immunoprecipitation where an antibody against a target protein precipitates an interacting protein. The efficiency of a particular method depends on the purity of tagged bait protein, the native state of interacting pair, and appropriate post-translational modification of prey protein to interact with bait, the concentration of bait protein and the interaction conditions (pH, buffer composition and cofactor requirement).

CONSENT FOR PUBLICATION

Not applicable

CONFLICT OF INTEREST

None Declare

ACKNOWLEDGEMENTS

I acknowledge the support of my colleagues at Govt Degree College Nawakadal especially the Librarian of the Institute (Ms. Asmat). She provided me the space where I could sit, concentrate and complete this portion of the book.

REFERENCES

[1]   Jones S, Thornton JM. Principles of protein-protein interactions. Proc Natl Acad Sci USA 1996; 93(1): 13-20.
       [http://dx.doi.org/10.1073/pnas.93.1.13] [PMID: 8552589]

[2]   Nooren IM, Thornton JM. Diversity of protein-protein interactions. EMBO J 2003; 22(14): 3486-92.
       [http://dx.doi.org/10.1093/emboj/cdg359] [PMID: 12853464]

[3]   Berggård T, Linse S, James P. Methods for the detection and analysis of protein-protein interactions. Proteomics 2007; 7(16): 2833-42.
       [http://dx.doi.org/10.1002/pmic.200700131] [PMID: 17640003]

[4]   Golemis E. Protein-protein interactions: a molecular cloning manual. New York: Cold Spring Harbor Press 2002; p. 682.

[5]   Bonifacino JS, Dell'Angelica EC, Springer TA. Immunoprecipitation. Curr Protoc Mol Biol 2001; Chapter 10: 16.
       [PMID: 18265056]

[6]   Phizicky EM, Fields S. Protein-protein interactions: methods for detection and analysis. Microbiol Rev 1995; 59(1): 94-123.
       [PMID: 7708014]

[7]   Einarson MB. Detection of Protein-Protein Interactions Using the GST Fusion Protein Pulldown Technique.Molecular Cloning: A Laboratory Manual. 3rd ed. Cold Spring Harbor Laboratory Press 2001; pp. 55-9.

[8]   Einarson MB, Orlinick JR. Identification of Protein-Protein Interactions with Glutathione S-Transferase Fusion Proteins.Protein-Protein Interactions: A Molecular Cloning Manual. Cold Spring Harbor Laboratory Press 2002; pp. 37-57.

[9]   Edmondson DG, Dent SYR. Identification of protein interactions by far western analysis. Curr Protoc Protein Sci 2001; Chapter 19: 7.
       [PMID: 18429148]

CHAPTER 7

Protein-Nucleic Acid Interactions

Abstract: The biological network is intricately woven by several entities which work in cellular system in a co-ordinated manner as a result of different types of interactions (protein-protein interactions, protein-RNA and protein-DNA interactions). Among them, the protein-nucleic interactions play an important role in regulating cell function and disruption among these interactions leads to catastrophic consequences within the biological system. The protein-nucleic acid interactions are integrated into several key cellular processes which include regulation of gene expression, replication, recombination, repair, translation, transcription, packaging of nucleic acids and the formation of cellular machinery. In this chapter, we will discuss the several methods by which we can analyse protein-nucleic acid interactions.

Keywords: Chromatin Immunoprecipitation, EMSA, Gel Shift Assays, miRNAs, SDS-PAGE, Transcriptional factors, Untranslated regions (UTR).

INTRODUCTION

Inside the cell, the extraction of information from DNA and timely utilization of the information requires interaction with proteins. The nucleic acid binding proteins have the ability to recognize either a specific sequence or secondary structure (major groove/minor groove) of the nucleic acid. The binding and manipulation of structures of DNA/RNA by proteins eventually play an important role in chromatin remodelling, synthesis of RNA and proteins and regulation of gene expression [1]. RNA-protein interactions are crucial not only for protein synthesis but in the regulatory roles of non-coding RNA. The non-coding RNAs include the microRNA's (miRNA), 5′ and 3′ untranslated regions (UTR) of mRNA and small interfering RNA (siRNA). They are highly significant entitities due to implication of their role in progression of various diseases, thereby urging a prompt need to understand protein-RNA interactions which regulates them [2].

The elucidation of protein-nucleic acid interactions is important for knowing wide range of cellular processes and understanding mechanisms underlying numerous diseases. The recent reports have suggested that many neurological disorders such as Alzheimer's, Huntington's, Parkinson's, and polyglutamine tract expansion diseases are a consequence of aberrant protein-DNA interactions that lead to changes in the normal patterns of gene expression [3]. Various techniques for deciphering protein-nucleic acid interactions and the locus of a DNA binding substrate protein of a nucleic acid are summarised below:

7.1. Electrophoretic Mobility Shift Assay (EMSA)

The DNA- protein interactions were primarily studied by nitrocellulose filter binding assay but were replaced by the concept of EMSA given by Fried and Crothers [4]. EMSA is one of the central techniques to understand gene regulation by determining the protein-DNA interactions. The technique is based on the fact that free DNA molecules migrate faster as compared to protein-DNA complexes when subjected to agarose or non-denaturing polyacrylamide gel electrophoresis. It is also known as gel-shift or gel-retardation assay because upon protein binding, the rate of DNA migration is either shifted or retarded.

Gel-shift assays are primarily used to identify sequence-specific DNA-binding protein (transcriptional factors) in crude lysates and coupled with mutagenesis; they are used to identify the important binding sequences within a given gene's upstream regulatory region. Transcription factors are regulatory proteins that form complexes with the specific DNA binding sequences and play an important role in transcription initiation. The transcriptional factors either act as an activator or repressor of expression of the targeted gene. They play a central and pivotal role during development and differentiation.

In a non-denaturing agarose/polyacrylamide gels, under the influence of electric field, the nucleic acids the nucleic acids being negatively charged will migrate towards the anode when electric current is applied. The migration of nucleic acids is impeded by the sieving effect of the gel in proportion to their molecular weight and charge. There is one more factor that affects migration of nucleic acid *i.e.* topology of nucleic acids. Due to confirmation and circularity of nucleic acid, the molecules seem longer or shorter than they really are [5]. When a protein is added in the mix that interacts with the nucleic acid forming complex with it, there is relative change in the gel migration compared to free nucleic acid as shown in Fig. (**7.1**). The shift occurs due to an increase in molecular weight, the alterations in charge and also changes in nucleic acid conformation.

The electrophoretic mobility shift assay has been also used to assess the different molecules based on thermodynamic and kinetic parameters. The stability of the complex during the brief time while passing through the gel depends upon the ability of the gel to resolve protein –DNA complexes. The relatively low ionic strength of the electrophoresis buffer stabilises the otherwise transient sequence specific interactions. When the complexes enter into the gel, it is stabilised by caging effects of the gel matrix. This implies that if the complex dissociates, its localized concentration remains high, leading to quick re-association. Thus the labile complexes can also be resolved by this method.

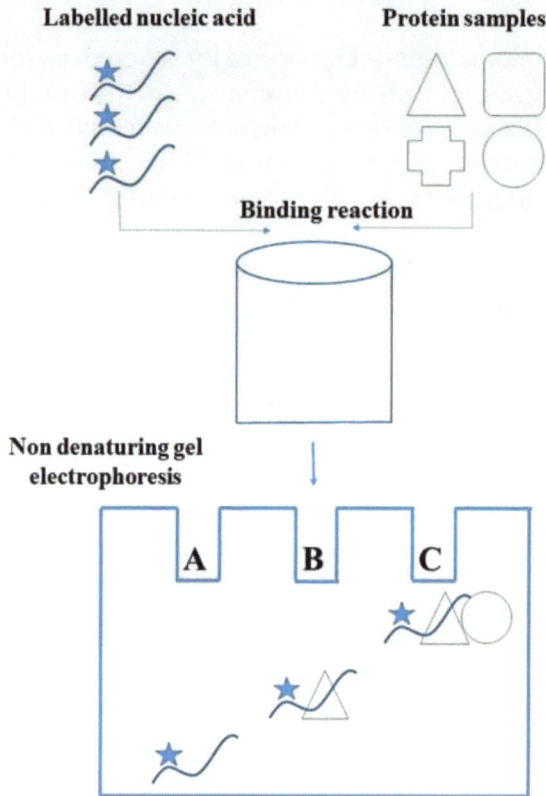

Fig. (7.1). Schematic Representation of EMSA. The lines with a star represent the labelled nucleic acids wheras the different shapes represent the different proteins. Both of them are mixed and allowed to undergo binding reaction and then loaded into a denaturing gel. The results are detected according to the label in the nucleic acid. The lane A represents a well in which labelled nucleic acids were loaded. The lane B represents the binding one protein to the nucleic acid. The lane C symbolises the binding of one nucleic acid to more than one protein. the free nucleic acid has higher mobility and moves faster than the bound molecules. The nucleic acid with more than one protein has the least mobility and is closer to the beginning of the gel.

The conventional EMSA protocol which is susceptible to optimization involved five different steps [6] which include:

Preparation of Protein Lysates: Depending upon whether the nucleic acid - interacting protein is already known or not, two different approaches are taken into consideration. The preparation of protein samples depends on which category it belongs so as to get optimal performance. The whole cell extracts are used while working on putative nucleic acid binding protein or complexes of unknown sub-cellular origin. The disadvantage of using whole cell extracts is that they are crude and are more susceptible to degradation due to presence of cellular proteases. To avoid this problem the protein samples are placed on low temperature and protease inhibitors are also included.

If the protein of interest is already known then we can prepare recombinant protein which can be easily expressed and purified. The recombinant proteins are usually expressed in either bacterial or eukaryotic cell expression systems. The purification of protein of interest is facilitated and assisted by using a tag. The commonly used tags for protein purification include glutathione-S-transferase (GST), tandem affinity purification tag (TAP tag), and maltose binding protein (MBP) or 6xHistidine. The gene of interest and tags are cloned in such a way that the fusion protein is expressed. Sometimes a cleavage site is placed in between the protein of interest and the tag so the latter can be easily removed after purification. .

Preparation and Labelling of Nucleic Acids: The synthesis and designing of the probe depends on the purpose and application of the study. The design of probe and its length will influence the detection and therefore the sensitivity of the results. The most common and sensitive method is radioisotope labelling of nucleic acids. The radioisotope (usually ^{32}P) is incorporated in the nucleic acid during its synthesis by the use of labelled nucleotides or afterwards *via* end labelling using a kinase or terminal transferase. Though the method is extremely sensitive but it may prove dangerous if extra safety measures are not available. The other techniques for nucleic acid labelling include use of fluorophores, biotin or dioxigenin. If these molecules are used for labelling, then the detection is achieved by chemilumniscence or immunohistochemistry. The unlabelled nucleic acids can also be used in a gel-retardation assay and can be detected by post electrophoretic staining using ethidium bromide or with fluorophores or chromophores that bind nucleic acids. However, the use of labelled nucleic acids is usually preferred as it more sensitive and facilitates easy detection.

The length of probe also depends upon the type of study. The small probes are used if we are looking for specific binding sites. They are easy to synthesise, cost effective and has less non-specific binding sites. In a short length probe, the binding sites are close to the molecular ends which can lead to uneven binding and the resolution of free nucleic acid from the complex will be difficult. On the other hand, longer sequences do not have such problems but they take more time to get resolved due to large size and there is non-specific binding.

Binding Reaction: The protein charge and conformation is sensitive to pH and salt concentration, thereby influencing the protein nucleic acid interactions. To maintain the favourable experimental conditions so that the data obtained from the *in vitro* conditions resemble that of *in vivo* conditions, different buffers are used. The most commonly used buffers include tris based buffers and 3-(N-morpholino)-propanesulfonic acid (MOPS), 4-(2-hydroxyethyl)-1-piperazinee-thanesulfonic acid (HEPES). The binding sometimes also requires the presence of co-factors or stabilising agents, so they are also added so to minimise non-specific binding. The addition of nuclease and phosphatise inhibitors avoids the degradation of protein samples.

To eliminate the unwanted loss of protein, the addition of a carrier protein like bovine serum albumin is incorporated. The addition of unlabelled competing nucleic acids is helpful when there are secondary binding activities that compete with the relevant one. This strategy works only if the protein of interest has higher affinity with the target nucleic acid. Since the presence of a competing nucleic acid will always minimise the amount of specific binding, checking different competitors and concentrations is needed to optimize the assay. One more option to overcome the problem of non-specific binding is the addition of salt. The salt at different concentrations has the ability to disrupt non-specific ionic bonds without affecting the specific interactions.

Non –Denaturing Gel Electrophoresis: The binding reaction is followed by non–denaturing gel electrophoresis and can be done on polyacrylamide or agarose gels. The selection of type of gel used depends on the size of the nucleic acid and the desired resolution of the gel [7]. The polyacrylamide gels are preferred over linear gels as there is gradient of pore sizes which are important to take into consideration when the complex has a much higher weight as compared to free nucleic acid. The polyacrylamide gels are preferred over linear gels as the gradient in pore size increases the range of molecular weight which is important when the complex has a much higher weight than the free nucleic acid. When the molecules have close molecular weight, the gradient gels are used which provide better resolution for nucleic acid protein complexes. The gradient gels are more likely to separate those with close molecular weight and provide better resolution for nucleic acid protein complexes. However, agarose gels are mostly used in assays where larger protein complexes or nucleic acids are studied.

For electrophoresis, care should be taken with respect to buffers ionic strength, pH, salt concentration so as to stabilise the protein complex generate less heat and to increase the speed of migration. The commonly used buffer for electrophoresis is Tris-Borate-EDTA, TAE *etc*. The temperature monitoring is important for agarose gels which could lead to dissociation of nucleic acid-protein complexes.

Detection: Depending upon the type of probe used, the detection of EMSA results is done. When the unlabelled nucleic acid is used, the shift in mobility is detected by staining with molecules that bind nucleic acids. But when the labelled nucleic acid is used, the detection methods depend upon the label of radio-isotope. The other labelling methods involve biotin, dioxigenin or fluorophores and detection is done by chemilumniscence or immunohistochemistry. These labels give less sensitive results and detection involves extra efforts like transfer to a membrane and incubation with primary and secondary antibodies *etc.*

In order to achieve specific results, the gel-retardation assay has been used under different conditions. This technique is used not only for studying protein-nucleic acid interactions but also for determining nucleic acid conformational characteristics. Earlier the measurement of kinetic and thermodynamic parameters was done by using EMSA. The association rates are studied by adding the complex components at known concentrations and loading them in a running gel after precise intervals. Whereas, the dissociation rates are studied by time course experiments by adding competing nucleic acids to the preformed complexes. The binding of protein to nucleic acids occurs in a cooperative behaviour which can be determined by the gel-retardation assay. For determining the stoichiometry of protein-nucleic acid complexes, more complex approach is used. After separation of free and complex nucleic acid on a non-denaturing gel, the proteins are treated with SDS and transferred to a membrane, followed by detection of protein using specific antibody. The relative stoichiometry is determined by studying protein bands as well as nucleic acids by autoradiography and finally quantified by densitometry [8].

7.2. Chromatin Immunoprecipitation (ChIP) Assay

ChIP is one of the important assays to identify links between genome and the proteome by monitoring transcriptional regulation through transcriptional factor-DNA binding interactions and histone modifications. It's a standard method for the elucidation of transcription factor binding sites. The assay is based on the ability to capture snapshots of specific protein-DNA interactions occurring inside the cells and then quantifying the interactions using quantitative Polymerase Chain Reaction [9].

In this assay, a cross-linking agent is used to covalently bind proteins and chromatin that are in direct contact. The cells are lysed, DNA is isolated and fragmented. The complex of proteins csross-linked with DNA is immunopre-cipitated by using specific antibodies against the target protein. The bound nucleic acid is released out by altering the cross-linking and then analysed. Previously, the analysis of DNA was done by PCR but it requires some previous knowledge of the candidate DNA regions. However, recently there are more sophisticated tools that allow millions of sequence reads. They are either coupled with micro-arrays (ChIP –chip) or high throughput sequencing (ChIP–seq) to get quick and reliable results.

7.3. Affinities Capture Methods

The coupling of nucleic acid sequencing and labelling technologies has provided platform for verifying and characterising the interaction of protein with particular nucleic acid sequence motifs. The nucleic acid sequences under study are labelled with amine or biotin tags linked to the 5'end *via* cross-linker [10]. These labels are then subjected to detection strategies that allow protein-nucleic acid interactions.

Plate Capture Method: There are different ways to immobilise DNA or RNA and then to analyse their interactions with specific protein. One of the methods involves use of 96 or 384 well microplates coated with streptavidin that bind biotinylated nucleic acids. The cellular lysates is prepared in binding buffer and allowed for sufficient time to allow putative binding protein to come in contact with immobilised oligonucleotide. After removal of extract, the wells are washed many times to remove non-specifically bound proteins. The detection is finally done by specific antibody labelled with alkaline phosphatise (ALP) or horse radish peroxidase (HRP).

7.4. Pull-Down Methods

The pull down method is another *in vitro* method for identifying the protein-nucleic acid interactions [11]. Here, like ELISA method, the biotin or amine labelled nucleic acid is immobilised on either streptavidin or amine-reactive gel surface. The gel is prepared according to the requirement of the experiment either in spin cup, column or batch format. First the nucleic acid bait is immobilised, cellular extract is prepared in binding buffer and incubated with each other. The incubation is allowed for sufficient time so that the putative protein binds the immobilised oligonucleotide. Then the gel is washed and the purified protein prey may be eluted by either salt gradient or changing buffer conditions that disrupt the interactions. The elution is followed by characterisation technique either SDS-PAGE or western blotting.

CONCLUDING REMARKS

The protein-nucleic acids interactions mediate wide range of processes including the maintenance of cellular metabolic and physiological balance. These specific interactions are crucial for regulation of DNA replication damage and repair, control over the process of transcription RNA processing and maturation, role in protein synthesis and transport of molecules across nucleus. The different genes are expressed at different levels due to binding of transcriptional factors to the promoter and regulatory regions of genes. The defects in this type of regulation have been implicated in numerous human diseases. Thus the ability to screen for transcriptional factors is important not only for gene regulation studies but also for drug discovery. There are different methods for studying protein-nucleic acid interactions like ELISA, chIP, affinity capture methods *etc*. The ELISA has the advantage to resolve the complexes of varying stoichiometric or conformation as compared to other assays. The advantage of studying DNA-protein interactions by an electrophoretic assay is the ability to resolve complexes of different stoichiometry or conformation. The source of DNA-binding protein may be a crude nuclear or whole cell extract rather than a purified preparation. However, several things should be kept in mind while studying protein nucleic acid interactions like:

1. Minimising the chances of protein and oligo degradation by incorporating proteases and nuclease inhibitors.
2. Avoid nonspecific binding of proteins by taking necessary measures.
3. Keeping the suitable conditions and addition of cofactors that are required for the proteins to bind the DNA or RNA. Also, some proteins may require the nucleic acid to be double or single stranded before binding can occur.
4. Care should be taken about the length of the carbon chain between label and the oligo. It can make a crucial difference by reducing the steric hindrance of the bound oligo.
5. For proper execution of the experiment, use of appropriate controls will be essential. The presence of control determines the success of the experiment.

CONSENT FOR PUBLICATION

Not applicable

CONFLICT OF INTEREST

None Declare

ACKNOWLEDGEMENTS

I express my gratitude to my friends especially Aneesa, Mehwish, Asmat, Humaira, Shaifta, Mansha for their unconditional support and timely help.

REFERENCES

[1]   Luger K, Phillips S. Protein-nucleic acid interactions. Curr Opin Struct Biol 2010; 20: 70-2.
[http://dx.doi.org/10.1016/j.sbi.2010.01.006] [PMID: 20133121]

[2] Bourne P, Murray-Rust J, Lakey JH. Protein-nucleic acid interactions. Folding and binding. Curr Opin Struct Biol 2001; 11(1): 9-10.
[http://dx.doi.org/10.1016/S0959-440X(00)00162-7] [PMID: 11179884]

[3] Cordeiro Y, Macedo B, Silva JL, Gomes MPB. Pathological implications of nucleic acid interactions with proteins associated with neurodegenerative diseases. Biophys Rev 2014; 6(1): 97-110.
[http://dx.doi.org/10.1007/s12551-013-0132-0] [PMID: 28509960]

[4] Fried M, Crothers DM. Equilibria and kinetics of lac repressor-operator interactions by polyacrylamide gel electrophoresis. Nucleic Acids Res 1981; 9(23): 6505-25.
[http://dx.doi.org/10.1093/nar/9.23.6505] [PMID: 6275366]

[5] Hellman LM, Fried MG. Electrophoretic mobility shift assay (EMSA) for detecting protein-nucleic acid interactions. Nat Protoc 2007; 2(8): 1849-61.
[http://dx.doi.org/10.1038/nprot.2007.249] [PMID: 17703195]

[6] Holden NS, Tacon CE. Principles and problems of the electrophoretic mobility shift assay. J Pharmacol Toxicol Methods 2011; 63(1): 7-14.
[http://dx.doi.org/10.1016/j.vascn.2010.03.002] [PMID: 20348003]

[7] Garner MM, Revzin A. A gel electrophoresis method for quantifying the binding of proteins to specific DNA regions: application to components of the Escherichia coli lactose operon regulatory system. Nucleic Acids Res 1981; 9(13): 3047-60.
[http://dx.doi.org/10.1093/nar/9.13.3047] [PMID: 6269071]

[8] Alves C, Cunha C. Electrophoretic Mobility Shift Assays Analyzing protein-nucleic acid interactions. Gel Electrophoresis-Advanced Techniques 2012; pp. 205-29.
[http://dx.doi.org/10.5772/37619]

[9] Massie CE, Mills IG. ChIPping away at gene regulation. EMBO Rep 2008; 9(4): 337-43.
[http://dx.doi.org/10.1038/embor.2008.44] [PMID: 18379585]

[10] Kadonaga JT, Tjian R. Affinity purification of sequence-specific DNA binding proteins. Proc Natl Acad Sci USA 1986; 83(16): 5889-93.
[http://dx.doi.org/10.1073/pnas.83.16.5889] [PMID: 3461465]

[11] Kneale G. DNA-protein interactions: principles and protocols. Methods Mol Biol 1994; 30: 1-20.
[PMID: 8004186]

GLOSSARY

Active site *Site at which substrate binds on an enzyme.*

Adaptor *A single stranded synthetic oligonucleotide that produces a molecule containing cohesive ends and a restriction site.*

Agarose gel Electrophoresis *It is a technique in which a matrix composed of a highly purified form of agar is used to separate large DNA and RNA molecules.*

Allele *One of the variant forms of a gene occurring at a given locus on a gene.*

Amino acids *The building blocks of proteins.*

Antibiotic *A chemical substance that is used to fight specific microbial infection.*

Antibody *A protein produced by the B lymphocytes as a consequence of exposure to a particular antigen and having the ability to specifically react with its complementary antigen.*

Antigen *A molecule responsible for eliciting immune response on entering inside an organism.*

Aseptic *Free from microorganisms.*

Assay *Test to detect presence of specific substances in small amounts in solution.*

Autoclave *Instrument used to sterilise glassware and culture media.*

Autoradiography *Technique that captures image formed in a photographic emulsion due to emission of radioactivity from a labelled component.*

Bacteriophage *Viruses that attacks and infects bacteria.*

Baculovirus *Virus that infects arthropods (mainly insects).*

Biolistics *A technique to introduce DNA inside a host cell.*

Blot transfer *Transfer of a blot from a gel to a membrane filter.*

Blunt end *The ends of a double stranded DNA that do not possess sticky end.*

Cell Suspension *Cells in liquid medium used often to describe suspension culture of a single cell and cell aggregates.*

Chromosomes *The threads of DNA in the nuclei that carry genetic inheritance.*

Clone *A collection of genetically identical cells or organisms that have identical genetic composition.*

Codon *A set of three nucleotides in mRNA that specify a tRNA carrying specific amino acid which is incorporated into a polypeptide chain.*

Contamination *Presence of unwanted microorganisms in microbial plant or animal cultures.*

Complementary DNA *DNA strand from messenger RNA using the enzyme reverse transcriptase.*

Cryopreservation *Preservation of culture at extremely low temperature of -196°C.*

Cryoprotectant *Chemical substance that inhibits the freezing and thawing damage to cells.*

de novo *A new, afresh.*

Denaturation *Separation of double stranded DNA into single strands.*

Dextran	*Polysaccharide produced by certain bacteria.*
Differentiation	*The biochemical and structural changes by which cells become specialized in a particular form and function.*
DNA Amplification	*Formation of many copies of DNA by using techniques like PCR.*
DNA polymerase	Enzyme *responsible for synthesising DNA.*
DNA probes	*Isolated single DNA strands used to detect the presence of the complementary (opposite) strands.*
Doubling time	*Time required doubling the number of cells.*
Electroporation	*Transitory opening of membrane pores by electrical pulses.*
Embryonic Stem Cells	*Embryo derived cells that give rise to all differentiated cells.*
Enzyme	*A class of protein that control biological reactions.*
Epitope	*Specific domain present on an antigen recognised by an antibody.*
Fibroblast	*Flattened Connective tissue.*
Gene	*A hereditary unit of DNA coding for a specific protein.*
Genome	*Total genetic material of an organism or individual.*
Haploid	*Cell with one set of chromosome.*
Immobilised Enzyme	*An enzyme that is bound or localised within a defined region allowing it to be reused in a continuous process.*
Immunoglobulins	*Antibodies produced by B-lymphocytes involved in humoral immunity.*
Klenow fragment	*Product of proteolytic digestion of DNA polymerase I.*
Ligase	*Enzymes used to join the segments of DNA.*
Microinjection	*Technique of injecting DNA or RNA into nucleus of cells using a micropipette.*
Monoclonal Antibodies	*Antibodies that recognises only one kind of antigen.*
Mutation	*Defects in the genetic material leading to stable changes on a gene inherited on reproduction.*
Oligonucleotide	*A short synthetic single stranded DNA.*
Oncogene	*Gene responsible for causing cancer.*
pH	*Measurement of acidity or alkalinity of a solution by knowing the negative logarithm of hydrogen ion concentration.*
Phenotype	*Physical characteristics of an organism*
Plasmid	*The extra nuclear DNA found in bacteria and some other organisms carrying the non-essential genes and replicates independent of the main DNA.*
Promoter Sequences	*A regulatory DNA sequence that initiates the expression of a gene.*
Protoplast	*Plant cell whose cell wall has been removed.*

Recombinant DNA *Hybrid DNA produced by joining pieces of DNA from different organisms.*

Restriction Enzymes *Enzymes that cleave DNA at specific sites.*

Scale up *Expansion of laboratory experiments to full sized industrial processes.*

Taq Polymerase *Heat stable DNA polymerase isolated from thermophilic bacteria (Thermus Aquaticus).*

Tissue culture *Growing individual cells or clumps of plant or animal tissue artificially.*

Transgene *Target gene responsible for development of transgenic organisms.*

Transient *Short duration.*

Vectors *Vehicles for transferring DNA from one cell to another.*

Virus *Infectious agent that need a host cell for replication.*

YAC *Yeast artificial chromosome used as DNA cloning vector.*

SUBJECT INDEX

C

Calcein 46, 47, 48
Carboxyfluorescein diacetatesuccinimidyl
    ester 44, 48
Carboxyl groups 59
Caspase activated DNase (CAD) 30
Caspase activation 35, 37
Caspase activity 37, 38, 39
Caspase activity assays 38, 39
   luminometric 38
Caspases 29, 30, 31, 32, 35, 37, 38, 39
   downstream 31
Caspase substrates 37
Catalase 58
Catalyses 58, 63
Cationic lipid 20, 23
C.elegans 30
Cell culture 1, 2, 3, 4, 5, 8, 9, 10, 11, 14, 15,
    16, 17, 18, 30, 50, 51, 54
   animal 2, 15
   assays 18
   conditions 51
   contaminants 16
   dishes 54
   establishing 15
   infected 17
   plastic ware 8, 9
   primary 6, 18
Cell death 17, 29, 30, 44, 51, 52, 64
   programmed 29
Cell division 1, 2, 24, 56
Cell extracts 72, 79, 83
Cell growth 2, 5, 10
Cell line authentication 1, 13, 14, 15
Cell lines 1, 3, 5, 6, 8, 9, 10, 11, 12, 13, 14, 15,
    16, 17, 18, 25, 26, 35
   contamination in 15, 16
   continuous 3, 6
   cryopreservation of 8, 11
   established 18
   finite 5, 6
   packaging 26
Cell lysates 69, 70
Cell membrane 22, 25, 29, 36, 37, 45

Cell monolayer 11
Cell population 5, 6, 46, 48
Cell proliferation 1, 44, 49, 56
Cell proliferation rate 48, 49
Cells 2, 3, 4, 6, 12, 13, 20, 24, 25, 27, 29, 30,
    32, 33, 35, 44, 47, 48, 49, 52, 53, 58, 72
   active 49
   adherent 3, 33, 35
   adjacent 30
   adjacent parenchymal 32
   bacterial 2, 72
   cryopreservant media 12
   cryoprotective agent's damage 13
   culture chick embryo heart 2
   cultured 24, 52
   embryonic 4
   fibroblastic 6
   haematopoietic 6
   healthy 33, 44
   hydrophobic 47
   lysed 52
   mammalian 20, 27
   microbial 2
   neoplastic 32
   non-apoptotic 35
   non-proliferating 48
   phagocytic 58
   regenerative 6
   sensitive 12
   target 25
   transformed 53
   tumour 29
Cells in dissociated cultures 5
Cells loosen 11
Cell strain 5
Cell suspension 11, 12, 25, 37
Cell thawing 13
Cell types 1, 2, 10, 13, 14, 21, 23, 25, 26, 33,
    35, 50
Cell viability 44, 48, 51, 56
   and cytotoxicity assays 44
Centrifuged cells 12
Chromatin condensation 29, 30, 32, 34
Chromatin Immunoprecipitation (ChIP) 76,
    81, 82
Chromophores 38, 39, 60, 79

www.ingramcontent.com/pod-product-compliance
Lightning Source LLC
Chambersburg PA
CBHW041720210326
41598CB00007B/722